RICCI'S SCIENTIFIC CONTRIBUTION TO CHINA

BY THE SAME AUTHOR

SAGESSE CHINOISE ET PHILOSOPHIE CHRÉTIENNE. Essai sur leurs relations historiques. *Tientsin*, 1935.

LA DÉCOUVERTE DE NESTORIENS MONGOLS AUX ORDOS ET L'HISTOIRE ANCIENNE DU CHRISTIANISME EN EXTRÊME-ORIENT. *Tientsin*, 1935.

AUX PORTES DE LA CHINE. Les missionnaires du XVIe siècle (1514-1588). *Tientsin*, 1933.

LE FRÈRE BENTO DE GOES. Chez les Musulmans de la Haute Asie (1603-1607). *Tientsin*, 1934.

AUX ORIGINES DU CIMETIÈRE DE CHALA. Le don princier de la Chine au Père Ricci (1610-1611). *Tientsin*, 1934.

IN PREPARATION

LE PÈRE MATTHIEU RICCI ET LA SOCIÉTÉ CHINOISE DE SON TEMPS (1583-1610).

LES ILES PHILIPPINES DU GRAND ARCHIPEL DE LA CHINE. Un essai de conquête spirituelle de l'Extrême-Orient.

L'HUMANISME CHRÉTIEN EN CHINE À LA FIN DES MING.

LES COMBATS DU PÈRE ADAM SCHALL POUR LA RÉFORME EUROPÉENNE DU CALENDRIER CHINOIS.

L'EMPEREUR K'ANG-HI À L'ÉCOLE DE LA CIVILISATION OCCIDENTALE.

Supplementary articles dealing with the religious and cultural history of the Far East:

Chrétiens nestoriens et missionnaires catholiques à la fin du XVIe siecle (*Bulletin catholique de Pékin*, 1932, t. XIX, pp. 176, 241).

L'Eglise catholique des XVIIe et XVIIIe siècles et sa place dans le développement de la civilisation chinoise (paraitra dans le premier numéro des *Monumenta Serica*).

L'art chrétien en Chine du temps du P. Matthieu Ricci (*Revue d'histoire des missions*, 1935, pp. 199-229).

La Mappemonde Ricci du Musée historique de Pékin (*Éditions de la Politique de Pékin*, 1928).

Les étapes de la cartographie scientifique de la Chine et des pays voisins (paraitra dans le second numéro des *Monumenta Serica*).

La musique européenne en Chine (*Bulletin catholique de Pékin*, 1935, pp. 40, 63).

Galilée et les Jésuites des Missions d'Orient (paraitra prochainement dans *La Revue des questions scientifiques*, Louvain).

Sur le Père Wieger, ont paru trois articles dans les *Digest of the Synodal Commission*:

1° Bibliographie méthodique des œuvres du P. L. Wieger (1932, t. V. pp. 975-984).

2° Missions, médecine et sinologie. Le P. L. Wieger (1934, t. VII, pp. 710-733).

3° Encore un mot au sujet du P. Wieger. A propos de ses « Controversiæ » (1934, t. VII, pp. 848-864).

CHRISTOPHORI
CLAVII BAMBERGENSIS
E SOCIETATE IESV.

ASTROLABIVM

CVM PRIVILEGIO

ROMAE,

Impensis Bartholomæi Graßi.

Ex Typographia Gabiana. M. D. XCIII.

SVPERIORVM PERMISSV.

P Mathao Riccio donum auctoris

alla China

ASTROLABIUM, BY CLAVIUS, ROME 1593

Title page reproduced from the copy presented to Ricci by the Author (see page 52, note 64).

Matteo Ricci's Scientific Contribution to China

By HENRI BERNARD, S.J.

Translated by

EDWARD CHALMERS WERNER

H.B.M. Consul, Foochow (retired),
Barrister-at-Law, Middle Temple,
Author of *Descriptive Sociology—Chinese*,
China of the Chinese, *Myths and Legends of China*, Etc.

HYPERION PRESS, INC.
WESTPORT, CONNECTICUT

Library of Congress Cataloging in Publication Data

Bernard, Henri.
Matteo Ricci's scientific contribution to China.

Translation of L'apport scientifique du père Matthieu Ricci à la Chine.
Reprint of the 1935 ed. published by H. Vetch, Peiping.
Bibliography: p.
1. Ricci, Matteo, 1552-1610. 2. Astronomy, Chinese. I. Title.
Q143.R5B43 1973 509'.2'4 73-863
ISBN 0-88355-059-8

Published in 1935
by Henri Vetch, Peiping, China

First Hyperion reprint edition 1973

Library of Congress Catalogue Number 73-863

ISBN 0-88355-059-8

Printed in the United States of America

PREFACE

Leibniz, in 1689, recommended to a missionary[1] "to promote the relations between Europe and China, which are like two worlds separated by an enormous distance: may they mutually instruct and enlighten each other!" In our day, the same wish is expressed in substantially identical forms[2]: "The endeavour of men of goodwill in our time has no other object than to bring about the coming of a humanism enlarged to the limits of the whole of Eurasia."

We know that, for about two centuries (from 1583 until the French Revolution), the Jesuits of China perseveringly attempted to establish this direct correspondence between the minds of the West and those of Farthest Asia, but the course of the intellectual exchanges did not always take the same direction. Whilst the monopoly of the "Portuguese Mission" (so called because only missionaries coming by way of Lisbon and Goa were admitted to Macao) lasted, the ideas generally travelled from the West to the East, and China was enriched by Europe's contribution more than Europe benefited by China's knowledge. After the advent of the "French Mission" (patronized by Louis XIV and his successors), China more and more shunned foreign influences whilst the savants of Europe

---

[1] Merkel, F. R., *G. W. von Leibniz und die China-Mission*, Leipzig, 1920, p. 31: "commercia inter duos velut orbes late divisos promovendi, commercia inquam doctrinæ et mutuæ lucis."

[2] Thus, by M. René Grousset, in the review *Europe nouvelle*, of December, 1932.

took a growing interest in the civilization of the Chinese, the interaction was principally from the Orient to the Occident.

The first man of modern times to inaugurate this "mutual commerce," was an Italian, the real founder of the "Portuguese Mission," Father Matteo Ricci. We have stated elsewhere what he owed to his predecessors (St. Francis Xavier, and the Fathers Valignano and Ruggieri).[3] From now on, without apprehension of resorting to technical language, we wish to indicate precisely what was his scientific contribution to China. Later, in a language accessible to the body of cultivated readers, we will study his profound action on the Chinese society of his time.[4]

NINGTSING,
OCTOBER 28, 1934.

[3] HENRI BERNARD, S.J., *Aux Portes de la Chine, Les missionnaires du XVIme siècle*, Tientsin, 1933.

[4] Until our full explanation of the apostolate methods of Father Matteo Ricci appears, our readers will find a first justification of his scientific activity in a chapter written by us on "L'Académie du Seigneur du Ciel" (Bernard, *Missionnaires*, pp. 229-244).

CONTENTS

Preface 1

Abridged Bibliography 5

Chap. I.—The Legacy of Islam in China and in Europe toward the end of the XVth century 7

Chap. II.—Ricci's Scientific Training 22

Chap. III.—Ricci and Chinese Science 37

Chap. IV.—The Problem of Chinese Astronomy .. 53

Chap. V.—The Solar Eclipse of December 15, 1610 73

Conclusion 92

Bibliography 94

Index 99

ILLUSTRATIONS

ASTROLABIUM, BY CLAVIUS, ROME 1593 ..	*Frontispiece*
EPITOME OF THE ALMAGEST OF PTOLEMY, 1564	*Facing page* 26
A PAGE FROM THE EPITOME WITH TWO ADDED DRAWINGS BY RICCI	„ 27
GEMMA FRISIUS, COSMOGRAPHICUS LIBER PETRI APIANI	„ 64
LEAF FROM THE PREFACE TO EUCLID WRITTEN BY RICCI IN 1607	„ 68
FIRST PAGE OF RICCI'S TRANSLATION OF EUCLID	„ 69

ACKNOWLEDGMENT

Permission to reproduce the first four illustrations taken from books in the Peit'ang Library, was granted by Bishop Paul Montaigne, Apostolic Vicar of Peking. The fifth and sixth illustrations come from the Chinese library of Dr. John C. Ferguson. For these courtesies the publisher makes his grateful acknowledgment.

Abridged Bibliography
of works most frequently cited.

BERNARD, GOES.—Henri Bernard, *Le F. Bento de Goes chez les Musulmans de la Haute Asie.* Tientsin, 1934.

BERNARD, MISSIONNAIRES.—Henri Bernard, *Aux Portes de la Chine. Les Missionnaires du XVIme siècle.* Tientsin, 1933.

GILES.—Lionel Giles, *Translations from the Chinese World-Map of Father Ricci, Geographical Journal,* 1918, vol. 52, pp. 367-385 ; 1919, vol. 53, pp. 19-30.

HAVRET.—*Mélanges sur la chronologie chinoise,* Shanghai, 1920.

LIBRARY.—Library of the old Chinese Jesuits, in Peking.

PFISTER.—*Notices biographiques et bibliographiques sur les jésuites de l'ancienne mission de Chine.* Shanghai, 1932-1934.

R.—Matteo Ricci, s.j., *Opere storiche,* edited by Father Tacchi Venturi. Macerata, 2 volumes, 1910-1913.

RIQUEBOURG.—De Riquebourg-Trigault, *Histoire de l'expédition chrétienne au royaume de la Chine entreprise par les Pères de la Compagnie de Jésus.* Lille, 1617.

SAUSSURE, ARCHIVES.—*Le zodiaque lunaire asiatique,* in the *Archives des sciences physiques et naturelles,* Neuchâtel, Switzerland, March, 1919.

SAUSSURE, ORIGINES.—*Les origines de l'astronomie chinoise* (photomechanic reproduction of articles which appeared in the *T'oung Pao,* 1909-1913, 1920-1921).

SMITH.—David Eugene Smith, *History of Mathematics.* 2 volumes. Boston, 1923.

SOUCIET.—*Observations mathématiques, astronomiques, chronologiques et physiques tirées des anciens livres chinois,* . . . Paris, 1732.

TRIGAULT.—*Litteræ Societatis Jesu e regno Sinarum,* 1610-1611. Augsbourg, 1615.

WIEGER.—*China Throughout the Ages.* Sienhsien, 1925.

WYLIE.—*Notes on Chinese Literature.* Shanghai, 1902.

CHAPTER I

The Legacy of Islam in China and in Europe toward the end of the 15th century

When one tries to discover the foundations and the direction of scientific thought in ancient China, one encounters an obstacle which is almost unknown in other countries : elsewhere, in fact, the greatest merit of a savant is his originality, the new path which he opens for the intellect ; here, what is especially important, is to travel in the old ruts and to repeat in a somewhat stereotyped form the aphorisms of antiquity[1] : in Arithmetic, for example, the most ambitious among the learned would only think of explaining the *Chiu-chang suan-shu*, 九章算術,[2] a method of calculation, which disappeared in 213 B.C. and was laboriously reconstructed by the scholars of the following centuries. Hence, a quasi-impossibility of dating the stopping-places of thought, and a wearisome monotony in the succession of commentaries or of explanations of commentaries. On this grey background, the 13th century stands out clearly ; mathematics progressed mightily, algebra was cultivated, astronomy flourished.[3] At the end of the 14th century, everything relapses gradually into secular sleep and, when Father Ricci arrived in the Far East (August 7, 1582), he only succeeded in discovering some not understood vestiges of a glorious past. In fact, the young missionary brought back to China the very science of which the scholars had lost the significance, but this science was rejuvenated, revivified, developed and completed, so effectually that "the scientific contribution of Father Ricci" may be

[1] Wylie, *Chinese Researches*, Shanghai 1897, pp. 174-175.

[2] Wieger, pp. 85, 457.

[3] Smith, I, pp. 268-269 ; II, p. 381.

considered as a resurrection, or rather a continuation of the best which the 13th century had produced. This curious fact can only be fully understood if, briefly sketching the parallel history of the sciences in China and in Europe, we draw up the inventory of "the legacy of Islam in those two parts of the world toward the end of the 15th century."

The historians of civilization have seen in Islam especially a barrier interposing itself between Europe and the Far East: on each side of the insurmountable obstacle, almost different humanities have evolved following independent formulas. To-day, we realize that the Mussulman world was not exclusively a watertight partition preventing direct exchanges between two civilizations, but that in many spheres, and especially in that of the intellect, it served partially as a link; it was through Islam that China became acquainted with Greek science, at the very time of the Mongol offensive which, in the 13th century, momentarily re-established the unity of Eurasia, and it was also through Islam, that Western Europe could obtain access to the source of the knowledge which the Greeks of the Orient were jealously monopolizing. Some characteristic incidents will suffice to accentuate this interdependence.

We are told of Gherardo of Cremona (in the 12th century)[4] that:

> from infancy, he had been brought up in the bosom of philosophy; he had learnt all its parts according to the teachings of the Latins; but love of the *Almagest* [the astronomical treatise of Ptolemy] which he could find nowhere among the Latins, led him to Toledo [to the Moors of Spain]. There, when he saw what an abundance of books had been written in the Arabic language on each subject, he felt pity for the poverty of books from which the Latins suffered, and in his desire to translate them, he learnt the Arabic language.... He devoted himself also to passing in review the whole of the Arabic literature, in the prudent fashion of a man walking in green pastures and who does not pluck all the flowers, but

[4] Cited by M. Doublet in his abridged *Histoire de l'astronomie*, Paris, 1922, pp. 148-150.

plaits a wreath of the best. He chose then, from each class of subjects, the books which seemed to him the most elegant; he began to translate them into the Latin language as his cherished inheritance, giving them the clearest and most intelligible form he could; and he did not stop, as long as his life lasted. At the age of 73 years, he entered on the road which all flesh must follow; it was in the year A.D. 1187.

At one stroke, seventy-four different works, written by Greeks or Arabs, became again accessible to the Latins; among others, a large part of the works of Aristotle, Archimedes, all of Euclid, treatises on medicine, and lastly the masterly work of Ptolemy (the *Almagest*). The impulse once given by Gherardo of Cremona, little by little the whole of the Arabic literature was translated into Latin.

About the same epoch, the Mussulmans of Persia rendered analogous services to the Chinese of the Far East: Yeh-lü Ch'u-ts'ai, 耶 律 楚 才, had established a large school in Peking, he accompanied Genghis Khan to Persia, and in 1210, borrowed from the Uighurs their calendar to adapt it to the Mongol empire.[5] The intellectual exchanges followed one after the other throughout the 13th century: in 1259, Hulagu, grandson of Genghis Khan, caused to be erected at Meragha, in the neighbourhood of Tabriz, an observatory provided with immense instruments, and he summoned to it distinguished astronomers of whom the most eminent, Nasir al-Din (1201-1274), philosopher, arithmetician, geometrician, is known especially by his astronomical tables, called *Ilkhanian*.

The Mongol dynasty of the Yüan in China willingly received these novelties: it favoured the free exchange of merchandise as well as of ideas, and the scientific development of the Far East culminated with the Chinese Kuo Shou-ching, 郭 守 敬, (1231-1316), a clever man of multiple aptitudes, who began by regularizing the rivers in the Peking region, and then distinguished himself as a mathematician and constructor of genius.[6] But, about the year 1360, the national culture again went back on itself: the Turco-

[5] WIEGER, p. 441.

[6] WIEGER, pp. 241, 349.

Mongols of the West were definitely converted with Tamerlane to the least hospitable Islamism, whilst the Mongols of the East found themselves driven from China by the revolt of the Mings (1368). We will confine ourselves to characterizing in a general manner the heritage which Islam left to China.[7]

We must not expect to find among the Arabs the same powerful genius, the same gift of scientific imagination, the same "enthusiasm," the same originality of thought that we have among the Greeks. The Arabs are before all else the pupils of the Greeks; their science is a continuation of Greek science which it preserves, cultivates, and on a number of important points develops and perfects......

Their object was to simplify and make lucid. Without having sufficient genius to make generalizations or any great synthesis, they are very good arrangers. They arrange logically. They classify and enumerate, and this simple gift of orderliness and lucidity is almost sufficient to explain the progress which they made. Their manner is didactic; they appear to address themselves not, like the Greeks, to some particular amateur or to some Mæcenas interested in learning for itself alone, but rather to all intelligent students. Their books remind one of some good secondary or university text-books.

If the Mussulmans gave way to the Greeks in researches in pure mathematics, they were their equals, if not superiors, in astronomical observations; their instruments, much larger, were constructed with greater care, and they were thus able to revise many of the essential data on the obliquity of the ecliptic, the precession of the equinoxes, the length of the year, the eccentricity of the solar orbit......To appreciate justly the Arabs' contribution to civilization, one must not rest content in comparing their works with the discoveries of the golden age of Greece, but rather remember that their contemporaries of Europe and the Far East would have continued to vegetate in almost complete ignorance of the exact sciences if they had not been guided and stimulated by them.

---

[7] Carra de Vaux, in *The Legacy of Islam*, edited by Arnold, Oxford, 1931, pp. 376-379.

When at Nanking and at Peking, Father Ricci was admitted to view the magnificent apparatuses which had been constructed by Kuo Shou-ching in the Mongol dynasty of the Yüan,[8] he could not repress a cry of admiration "he had seen nothing better in Europe."[9] The genial Tycho Brahe (1546-1601) had just revolutionized the technique of astronomical observations in his Uraniborg observatory, but, when Ricci was on the point of leaving Italy and Portugal (1577-1578), the young savant was only just beginning his career in Denmark; the quadrant of the circle of nearly six metres in radius which had been destroyed by a tempest in 1574 had been made entirely of oak, and the construction of immense apparatuses in copper which astonished the cultivated minds of the Occident was still later. The official history of the Mongol dynasty of the Yüan has preserved for us a long description of the instruments of Kuo Shou-ching; at Nanking Ricci noticed especially four which he describes pretty minutely; later he saw those at Peking which are less carefully modelled. The historians of astronomy will make profitable use of what he said about them to solve certain enigmas; let it suffice for us to enlighten his text by some explanations borrowed from the History of the Yüan.

The first apparatus is the *Hun-hsiang*, 渾象, or "Celestial Sphere." It is made round like a ball, six feet in diameter, the degrees being marked both longitudinally and latitudinally. The equator is in the centre, equidistant from the two poles, in each case a quarter of the circumference. The ecliptic is elevated above and depressed below the equator, in each case barely 24 degrees. The elevations and depressions of the moon in its orbit being variable a bamboo hoop divided into degrees equally throughout is used to verify the intersections with the ecliptic, and moved from time to time accordingly. By observations first taken by the Compendium Instrument (the fourth one described hereunder) the right ascension and polar distance is ascertained, and according to these numbers the hoop being placed on the globe the proximity and angle of the moon's

[8] WYLIE, *Chinese Researches*, pp. 4 *seq.*

[9] R, I, pp. 315-317. The citations from the History of the Yüan (*Yüan Shih*) are taken from the translation by WYLIE, *Chinese Researches*, pp. 4 *seq.*

orbit with the ecliptic and equator is very obvious and easily determined when the theory of the movement is developed by calculation. The globe is placed on a square box, the north and south poles being respectively above and below the surface full forty degrees, half the globe being visible and half concealed. There are toothed wheels set in motion by machinery concealed within the box for turning the globe.

A second instrument (*Ling-lung i*, 玲瓏儀, armillary sphere) was a great sphere, not less in diameter than that measure of the outstretched arms which is commonly called a geometric pace. It has a horizon and poles; instead of circles it was provided with certain double hoops (*armillæ*), the void space between the pair serving the purpose of the circles of our spheres. All these were divided into 365 degrees and some odd minutes. There was no globe to represent the earth in the centre, but there was a certain tube, bored like a gun-barrel, which could readily be turned about and fixed to any azimuth or any altitude so as to observe any particular star through the tube, just as we do with our vane-sights;—not at all a despicable device! . . . (A similar instrument to that described by Father Ricci is still to be seen at the present day in the Central Observatory at Peking).

The third machine was a gnomon (*Kao-piao*, 高表; the history of the Yüan describes a similar one of much larger dimensions); the height of which was twice the diameter of the former instrument, erected on a very large and long slab of marble, on the northern side of the terrace. The stone slab had a channel cut round the margin, to be filled with water in order to determine whether the slab was level or not, and the style was set vertical as in hour-dials. We may suppose this gnomon to have been erected that by its aid the shadow at the solstices and equinoxes might be precisely noted, for in that view both the slab and the style were graduated.

The fourth and last instrument (*Chien-i*, 簡儀, "Compendium Instrument"), a specimen of which still exists at Peking and which is described at great length in the History of the Yüan), was one consisting as it were of three or four huge astrolabes in juxtaposition; each of them having a diameter of such a geometrical pace as I have specified. The fiducial line, or *Alhidada*, as it is called, was not lacking, nor yet the *Dioptra*. Of these astrolabes, one having a tilted position in the direction of the south, represented the equator; a second, which stood crosswise on the first, in a north and south plane, the Father took for a meridian; but it could be turned round on its axis; a third stood in the meridian plane with its axis perpendicular, and seemed to stand for a vertical circle; but this could also be turned round so as to show any vertical whatever. Moreover all these were graduated and the degrees marked by prominent studs of iron, so that in the night the graduations could be read by the touch without a light. All this compound astrolabe instrument was erected on a level marble platform with channels round it for levelling.

On each of these instruments explanations of everything were given in Chinese characters; and there were also engraved the 24 zodiacal constellations which answer to our 12 signs, two to each. (With regard to the last instrument, the history of the Yüan remarks that all the parts, which are not directly utilized for making observations, are decorated with clouds and that the pillars are in the form of dragons).

At the present day, there may still be seen in the lower court of the Central Observatory at Peking, analogous instruments, the second and the fourth apparatuses, dating from the Mongol dynasty; they are still more instructive in that, by mounting the steps leading to a platform raised above the ramparts of the Tartar city, one can compare them with the splendid instruments which Father Verbiest constructed[10] following the traditions of Tycho Brahe, about 1669-1670, sixty years after the death of Ricci. They are a most eloquent witness of the relative perfection to which the astronomy of the Mussulmans in the Far East had already attained.

There was, however, one common error to all the instruments, viz., that, in all, the elevation of the pole was assumed to be 36°. Now there can be no question about the fact that the city of Nanking lies in lat. 32¼°; whence it would seem probable that these instruments were made for another locality and had been erected at Nanking, without reference to its position, by some one ill-versed in mathematical science. [Wylie remarks lower down: "The first explanations of this puzzle I received in a hint from Col. Yule, who drew my attention to the fact that Yelu Chutsai (Yeh-lü ch'u-ts'ai) had

[10] H. Bosmans, *Ferdinand Verbiest, Directeur, de l'observatoire de Pékin* (1623-1688) (extract from the *Revue des Questions scientifiques*, January-April, 1912, Louvain, 1912, pp. 76 *seq.*)

established a college for the Mongol youth at P'ing-yang (平陽) in Shansi, which is actually in latitude 36° 6'."]

At the end of the 16th century, incomprehension had become such that no one was any longer capable of explaining their use. The official history of the Ming dynasty gives us the reason for this complete eclipse: on the fall of the Mongols, they began (1364) by substituting for the Yüan calendar ephemerides calculated according to the ancient Chinese rules (*Ta t'ung-li*, 大統曆);[11] then, after four years,—probably because the predictions of the official astronomers were too defective,—they re-established a Bureau of Mussulmans which also computed its own calendar (*Hui-hui li*, 回回曆). In 1374, there were discovered in the Palace Library[12] some hundreds of Persian manuscripts, and the emperor ordered them to be translated into Chinese; they did not find that they were erudite enough to understand the technical expressions and the scientific demonstrations, but contented themselves with utilizing the tables of practical astronomy, all the rest remaining unpublished ended by disappearing, especially the theory of the planets and some other books of theoretical mathematics[13]. Even the celestial observations of Tamerlane's grandson, the Persian Ulugh Beg (1393-1449), appear not to have been known in China. Thus the Legacy of Islam, in astronomy, was almost entirely dissipated.

Parallel to this stagnation followed by a retrogression in the Far East, it would be interesting to see how Europe quickened its step and, from certain points of view, succeeded in outdistancing its masters.[14] Under the

---

[11] MIKAMI, pp. 106-107.—WIEGER, p. 249.

[12] SOUCIET, II, p. 116.

[13] TRIGAULT, p. 127.—M. PELLIOT wrote in the *T'oung Pao*, 1934, vol. 31, p. 185: "I doubt as to the number (of the Persian manuscripts), and the only translations of which any trace has been found are: a calendar (re-issued), a Materia Medica (incorporated in the *Yung Lo ta-tien*, but lost)." The testimony of Father Trigault, who in 1613 wrote the annual Letter, demonstrates that more faith may be placed in the text of Father Souciet.

[14] For the summary account of the state of the sciences in Europe, we have frequently made use, sometimes quoting from it literally, of the substantial summary of Father LAURAND, *Manuel des études grecques et latines*, Paris, 1930, tome 3, Appendice I: Les Sciences dans l'antiquité.

domination of the Spanish sovereign Alfonso X the Wise (1223-1284), the collaboration of the Mussulmans and the Christians was affirmed with rapture in the astronomical tables, called *Alfonsine Tables,* the use of which was imposed until the beginning of the 16th century. An elementary treatise (*Sphæra* or *Sphæricum opusculum*) had a still more brilliant fortune; its author was an Englishman of the 13th century, a native of Hollywood whose name was latinized as Johannes Sacrobosco, "The Sacred Wood": the "Sphere of Sacrobosco," of indifferent value enough, reigned in the schools until the beginning of the 17th century and its fortune did not begin to decline until the day when the German George Purbach (1436-1476) had composed his "Theories of the Planets." The decisive event, which allowed the Latin scholars to break loose from the Arab translations and return to the original Greek ones, was the taking of Constantinople by the Turks (1453), followed by the exodus of the ancient manuscripts toward Western Europe.

The principal interest of this astronomy lay in its astrological applications, and in that respect the West was scarcely more reasonable than the East: the mute language of the skies allowed, it was believed, of prognosticating the future or of explaining the past. "The Philosophers," Kepler, himself an astrologer, said later on, "the Philosophers, whilst priding themselves on their wisdom, should not blame with so much bitterness the daughter of Astronomy: it was this daughter who nourished her mother. How small, in fact, would be the number of learned men who would have devoted themselves to Astronomy, if men had not hoped to read future events in the sky."

Very fortunately for pure science, the *Almagest* of Ptolemy (3rd century A.D.) had preserved the results acquired by the ancients: about a thousand fixed stars had been counted, and, if this number is insignificant in comparison with what has been discovered at the present day, at least that science had developed greatly; all educated people could tell the constellations of the sky, thanks to the tables and especially to the "spheres" on which they were depicted. They enumerated six, or rather seven planets (with the Sun, which they regarded as revolving round the earth), the Moon, Mercury, Venus, Mars, Jupiter, Saturn, and they knew that "the Morning Star" and "the Evening Star" were the same planet, Venus. Thanks to what was known of the apparent movements of the sun, of the moon, and of the planets, they could predict eclipses and they knew the explanation of them: an eclipse of the sun, when the moon is between the sun and the earth; an eclipse of the moon, when the earth is between the sun and the moon.

The number of astronomical observations continued to multiply, but there had not yet been invented either the telescope or the astronomical spy-glass, nor even instruments exact enough to calculate angles or to measure time. Nowhere was this defect in exactness felt more poignantly than in determining the different places of the earth on the geographical maps: latitude, and especially longitude, was only obtained with the enormous errors of 10°, 20° and sometimes 40°! Ptolemy was still for them the great initiator, but those who belonged to his school often only succeeded in representing the continents with monstrous deformations. The Chinese, minute and honest observers, ignored these problems whose solution is essential to scientific cartography, and contented themselves with accumulating valuable information without succeeding in arranging it in a satisfactory manner.

What we have just said about astronomy and its applications may be repeated with regard to the other sciences. Natural History was well enough known in Europe and even in China. Aristotle had been, in the West, the creator of scientific zoology and of comparative anatomy; he had collected an immense number of facts, the sum total of errors in which were of little importance in comparison with the truths he had observed. His successors, Pliny among others, had been much less well inspired; they constantly had Aristotle before their eyes, often they confined themselves to recapitulating him, but sometimes they added assertions unworthy of any credence or even absolutely ridiculous fables. It has been said: "Aristotle assembled facts in order to draw ideas from them, Pliny confined himself to collecting facts, and took them from wherever he could find them, except perhaps from nature." The Arabs resembled Pliny in their care in collecting facts from everywhere. The Chinese were rather related to the great Aristotle by their quest for exactitude; this last quality manifests itself especially in their immense treatises on Botany, drawn up especially from the medical point of view and somewhat similar to the compilations in which Dioscorides (1st century A.D.) had condensed the knowledge of the Greek naturalists.

Of Physics, understood in the modern sense and not as a part of Philosophy, Europe knew little, though she had a slight knowledge of Optics; China knew still less. The majority of the theories by which attempts were made to explain the phenomena of nature were more closely related to philosophy than to physics: in particular the Chinese theory of the five elements was of scarcely any interest for the history of the sciences. By contrast, the hypothesis of the Greeks and the Arabs on the subject of the four elements contained some useful ideas. All bodies are solid, liquid and gaseous; and they can assume an incandescent state. This fact of direct observation is the foundation of the theory: solids are called "earth"; liquids, "water"; gas, "air"; incandescent bodies, "fire." Elsewhere it had been noticed that fire is never alone, that it is always in one of the three other elements. Many European texts, dealing with the four elements, can be reasonably interpreted in the sense of "solid, liquid, gaseous, incandescent"; but it is otherwise with the musings concerning the Chinese five elements, and, even in Europe, there are to be found quite a number of speculations on the nature of bodies that all interpretation, however benign, fails to vindicate.

In short, it was in the domain of Pure Mathematics that science had up till then greatly developed, but whilst the powerful current of Muslim thought was progressively growing in Europe and had there spread very widely, it was seen to contract in China and to be confined to a minute field.

The ARITHMETIC of the Arabs or the art of reckoning had been a much humbler discipline than that of the Greeks or the science of numbers (the latter called Logistics, was allied rather with what Leibniz and other superior minds believed they had discovered in the ancient study of Mutations, so dear to the Chinese of all times). In China as in Europe, the reckonings of bankers and still more those of the State absolutely required good calculators. The invention of the figures known as "arabic" greatly simplified the systems of notations: besides the four operations, the extraction of square roots and ordinary fractions were known in Europe; decimal fractions were still unknown, and even ordinary fractions were only employed for very simple numbers. If it is true that the principle of positive and negative numbers had been discovered there, it was scarcely known how to apply them. China was much less advanced; they did not write the numbers, they represented them materially with the help of the abacus, an instrument composed of several rods on which balls were threaded; to work a calculation, one "played" (and one still "plays" at the present day) the operations by causing the balls to slide along the rods.

Similarly, GEOMETRY was, in China, intuitive: they did not justify its theorems by abstract demonstrations, but confined themselves to numerical measures. The Arabs had transmitted to Europe the wonderful "Elements of Euclid." The cause of that work's prodigious success, was not the originality of the conceptions: hardly anything, it seems, in it was really new in its details, but it gave a synthesis of Geometry so exact and so clear that it had supplanted all analogous treatises. The mind, in fact, passed in it without effort, by the smoothest descent, from one truth to another, and a whole immense doctrine unfolded itself in it with perfect regularity, without a gap and without a shock. Beside these "Elements," the geometry of the Greeks and the Arabs comprised the work of Apollonius on conic sections and other small works in which they succeeded in treating by geometrical methods questions the solution of which is at the present day furnished by algebra, like the equations of the second degree. For example, instead of saying $(a+b)^2=a^2+b^2+2ab$, they proved that the square constructed geometrically on a line formed of two lengths placed side by side a and b, is equal to a square whose side is a, plus another square whose side is b, plus twice a rectangle having for its sides a and b. The expressions "square number,"...... are actually a souvenir of this time when the geometrical conception ruled even in arithmetic. If, by algebra, one understands not only the science which permits one to solve the equation $ax^2+bx+c=0$ expressed by these symbols, but also a science which employs other signs (more or less suitable elsewhere), one may say that it had been virtually discovered from the 3rd century A.D., and that the Arabs had revived the knowledge of it;[15] under the Mongol dynasty in the 13th century, China knew how to calculate what corresponded to the equations of the 3rd and 4th degrees[16] and she applied herself to resolving the problem of squaring the circle.

As everywhere else, the momentary advance had been followed by a profound retrogression. The Persian treatises on trigonometry, and especially spherical trigonometry, had remained a dead letter for the scholars after the expulsion of the Mongols, and the official education more and more neglected the exact sciences. In the programme that the founder of the Ming dynasty ordained for the educational establishments (1370), were mentioned[17] "instruction in the rites, music, archery, horsemanship, writing, calculation and mathematics." Twelve years later (1382), they particularized the following too brief summary.

---

[15] Smith, II, p. 378.

[16] Smith, II, p. 380.

[17] Biot, *Essai sur l'histoire de l'instruction publique en Chine*, 1847, pp. 422 *seq.*

1. Each pupil received a copy of the following works: *Ching* (five canonical books), histories, collection of laws and edicts, books on the rites and external deportment.

All are to study these for the examinations.

2. On the first and the fifteenth days of the moon, they are to practise archery . . . following the ancient ritual.

3. The pupils are to practise writing, conforming to the principles of the ancient calligraphists. Each day, they are to furnish 500 characters.

4. Finally, for the study of calculation and of mathematics, they are to solve problems extracted from the *Chiu-chang suan-shu*.

Thus then, they completely ignored the contribution of the Mussulmans and reverted to the rudimentary arithmetic of the 3rd century B.C.! Theoretically, the candidates for the competitive examinations had to study a little mathematics; in practice, they showed little eagerness in that direction, and still less for the military exercises: they wished to occupy themselves solely with literature. They were allowed to do so. From the year 1384, there were only three tests for the competitions for admission to administrative posts: the first test related to the four classical books and the Five *Ching*, the second to imperial decisions and ordinances, the third to questions relating to the *Ching*, the historians, and contemporary affairs. From that time, mathematics, excluded from the official programmes, was no longer cultivated except by a few rare scholars.

A missionary of the end of the 17th century has well described what was then the ideal of high culture among the Chinese: [18]

The great and only Road to Riches, Honour, and Employments, is the Study of the *King* (or canonical Books), History, the Laws and Morality; also to learn to do what they call *Wên chang*, that is, to write in a polite Manner, in Terms well chosen, and suitable to the Subject treated upon. By this Means they become Doctors, and that Degree is once obtained, they are possessed of such Honour and Credit, that the

[18] Du Halde: *A Description of the Empire of China......*, London, 1741. Vol. II, p. 124.

Conveniences of Life follow soon after, because then they are sure to have a Government in a short time: Even those who return into their Provinces to wait for Posts, are in great Consideration with the Mandarin of the Place; they protect their Families against all Vexations, and there enjoy a great many Privileges. But as nothing like this is to be hoped for by those who apply themselves to the speculative Sciences, and as the Study of them is not the Road to Honours and Riches, it is no wonder that those sorts of abstracted Sciences should be neglected by the *Chinese*.

To sum up, thanks to the influence of Islam, mathematics, astronomy and perhaps medical botany had formerly reached a fairly high degree of perfection in China but, toward the end of the 15th century, the Chinese had rejected almost all of what they had been able to collect of this inheritance and the Mussulman colonies which had established themselves in various parts of the empire exercised no intellectual influence whatever. On the contrary, Western Europe having almost entirely succeeded in assimilating what Islam had bequeathed to it, began to progress through its own talents: the Mussulmans gradually disappeared from its horizon, but the track of their works did not cease to be recognizable in the continually rising tide of discoveries.

The sciences most in demand . . . . are a perfect knowledge of the language, laws, history and moral philosophy, because these are the roads by which they can reach the highest official posts. No one can be received into the class of doctors if he does not well understand the language, if he does not write the characters perfectly, and if he is incapable of composing an elegant discourse on the principal maxims of their ethics and government which are always drawn from what are called the canonical books.

For a candidate, the essential was to acquire knowledge by which " one could advance and become wealthy," and, " in view of the scant attention that was always given to rewarding those who excelled in the abstract sciences," one need not be astonished that the latter were neglected.

CHAPTER II

Ricci's Scientific Training

While the Chinese mind was satisfied with searching in the classical books for the principle of all wisdom, Europe from the end of the 15th century threw herself as if at play into intellectual adventures of all kinds. The First Renaissance,—almost up to the middle of the 16th century,—had nearly lost its way in fruitless reveries; the positive results were drowned in a current of daring speculations, renewed from neo-platonism or systems anterior to Aristotle. Toward the end of the Italian wars, political circumstances brought about, both in Protestant countries and in Catholic regions, a marked reaction against the innovators: the authority of the traditional teaching, which had seemed singularly shaken, consolidated itself to such a point that certain universities, far from favouring the necessary reform, be it what it might, were thenceforward the principal obstacle to be overcome by the scientific current. An advantage resulted from this reaction: thenceforth, cold reason and the critical observation of facts remained alone in the presence of the worm-eaten edifice of the science of the Middle Age. Father Ricci belongs to that period in which modern thought begins to extricate itself with definite characteristics of its own.

At that time nothing was known of a conflict, often sharp in modern times, between the "sciences" and "letters": both had the same friends. These friends had been especially the physicians, in the first half of the 16th century; now, it began to be the fashion to study law to finish one's education. This, from several points of view, was a great step in advance: the aphorism of Hippocrates: "Art is long, life is short, experience is dangerous, reason is uncertain," had, in fact, reigned too long; in contradiction of this axiom, the jurist did not wish to form any convictions except on clear reasonings and precise facts. Where the physician was obliged to come to an immediate decision in an urgent case, the jurist could retrench himself behind a *non liquet* and demand fuller information; thus the true scientific method gradually disengaged itself from the medley of blindly accepted traditions and hastily interpreted observations.[1]

Ricci, after a good study of the classic humanities at the college of Macerata, had applied himself for almost three years to the juridical sciences at Rome (October or November 1567-August 1571) before entering the Society of Jesus: it was especially as a young religionist during his sojourn at the Collegio Romano (September 1572-May 18, 1578) that he was able to initiate himself into the exact sciences as they had been taught for some years in the quite recent Order of the Jesuits.

It will be of use to draw in its main features a picture of the scientific achievements in Europe about 1575-1580; on this background, the pedagogic tradition of the Collegio Romano will stand out in great relief, and then some ascertained facts will serve to reveal the definitive acquisitions of the young Father Ricci.

The Europe of the end of the 16th century[2] had not yet seen the building-up of the Natural Sciences as they are conceived in our modern classifications. Botany, Zoology, Anatomy, Physiology, Geology and, we may add, Chemistry too often confused with Alchemy, had hardly emerged from the indistinct state in which the Arabs had left them.

On the other hand, in Pure Mathematics, appreciable progress had been made. In Geometry, the results obtained by Euclid were generally known and the elementary properties of conic sections were understood, but the scientific level remained lower than that of the ancient Greeks. The methods of Arithmetic became simpler day by day. Algebra was groping in search of

---

[1] TANNERY, in *l'Histoire générale* of Lavisse and Rambaud, vol. v, pp. 450 *sq.*

[2] ROUSE BALL, in *The Cambridge Modern History*, 1926, vol. v, pp. 707-708, 723.

really adequate symbolic notations. The simplest propositions of Trigonometry were easily accessible. This seems little, but that little sufficed to admit of approach to new problems, and in certain countries, such as Italy, the researches were pursued with an incredible ardour: and fortunate discoveries crowned the efforts of a Cardan or a Tartaglia.

In other fields, the advance was much less pronounced. Applied Mathematics remained in a pretty miserable state. Among these, the mechanism of rigid bodies was the oldest; Statics, as far as concerns parallel forces, had been placed on a solid base by Archimedes who had enunciated the fundamental principle: "When two equal weights are suspended from a rigid beam and at an equal distance on both sides, they are in equilibrium." But the problem of the resultant of forces acting on a particle, except in the special case of two rectangular forces, was still unknown. Dynamics, as a science, did not exist; it is true that it was stated on the authority of Aristotle that the velocity of the fall of heavy bodies varied proportionately to their weight,—and this principle which cannot have held a single instant in experience, was blindly accepted; and moreover on this false premise, they had not constructed any theory. In Astronomy, the authority of Ptolemy was almost as undisputed as that of Aristotle in the sciences, although the explanation of the movement of the planets by epicycles and eccentrics met with increasing difficulties as observations multiplied. The other subjects to which Mathematics were applied were especially Optics and Hydrostatics. In Optics, the law of reflexion was known, and they could solve without difficulty the most elementary geometrical problems in the matter of rays reflected by spherical surfaces. In Hydrostatics, the theory of floating bodies, enunciated by Archimedes, was fairly well known among the learned. As to the other branches of Physics, such as the study of Sound and of Electricity, it can be said that the little that was known does not merit description; in these matters, the authority of Aristotle remained supreme.

It may easily be conceived that for this science, still in its infancy, the knowledge required could be held in the hollow of the hand; and when one wished to devote oneself to personal researches of a special kind, it sufficed to know enough Latin to read the works of the ancients and the moderns. From the 15th century, friends of scientific culture had begun to found, in several Italian cities, groups or societies; they gathered together under the patronage of a great lord or a powerful personage, often in his own home, and there discussed the most diverse problems. A society of this kind, at Florence under Cosimo de' Medici, devoted itself to the study of Plato's writings, and, in order to emphasize this intellectual dependence, took the title of Academy. In the course of time, all groups of the same kind took to themselves the same name of Academy.

In the scientific movement of the 16th century, the Society of Jesus occupied quite a special place. Its founder, St. Ignatius of Loyola, after having witnessed in Spain the blooming of a literary humanism, inspired especially by Erasmus, had dwelt for a long while in Paris amid the fiercest struggles between the partisans and the adversaries of the new ideas. At Montaigu, he had experienced the austere rule of Noël Beda, the implacable enemy of the methods of the Renaissance, and at the college of Sainte Barbe, he had lived quite near to Calvin, the rigid reformist. The first preoccupations and the early cares of his apostleship in Italy turned him and his nascent Society toward the pilgrimage to Jerusalem, the preaching of the meek, secluded retreats, ... but soon he was led to organize colleges to train recruits for his Order and students from outside.

The first foundation was at Messina and was entrusted to a graduate of the University of Paris, Father Nadal. He was a scholar after the fashion of his time, knowing Greek and Hebrew; he had associated in Paris with the representatives of the "heroic" age of the Renaissance, amongst others Guillaume Postel, an extravagant enlightener and prophet who, having spent some time in the

Society of Jesus, had to be excluded from it because of his astrological predictions. In the programme of studies, Nadal was not inclined to exclude the sciences,[3] but he considered them as a rather accessory matter: "first some books of Euclid, he said, so that one may habituate oneself to demonstrations;[4] then the practical Arithmetic of Oronce [Finé] and his Sphere,[5] the Astrolabe of Stöffler[6] and Purbach's[7] Theories [of the Planets]."

Another plan also drawn up by Nadal for the Collegio Romano[8] details these essential arrangements for the use of the young philosophers of the Society of Jesus, with the hope that in three school years, at the rate of three-quarters of an hour a day, they will have learnt "at least the principles of the whole of mathematics:"[9]

1st year.—Euclid, a practical arithmetic, the principles of astronomy called the Sphere (the Cosmography of Oronce, if there is no more suitable one) . . . Granted that the practice of arithmetic and of the Sphere would ordinarily not take any considerable time, Euclid is to be explained as much as possible. Perhaps it will be advantageous to give demonstrations according to other mathematicians, such as the "Triangles of all kinds" of Regiomontanus[10] or [the works of] Jordanus [Le Forestier][11]

2nd year.—The Theory of Music and perspective (optics). The former will be taught either by following the common

---

[3] *Monumenta Pædagogica*, Madrid 1901, p. 621: regulations drawn up before 1553 (in the collection *Monumenta Historica Societatis Jesu*).

[4] Euclid was printed for the first time in 1482.—Cf. *Elementum Geometricum ex Euclidis Geometriæ*, JEAN VOEGELIN, Paris 1543 (LIBRARY).

[5] Oronce Finé was a pretentious compiler (SMITH, I, p. 308): on his *Pratica Arithmetica*, cf. SMITH, *Rara Arithmetica*, 1908, pp. 160 *seq.*

[6] JOANNES STOEFFLERINUS, *Elucidatio fabricæ ususque Astrolabii*, Paris 1553 (LIBRARY).

[7] *Theoricæ novæ planetarum Georgii Purbachii......*, Paris 1543, Basel 1573, Cologne 1581 and 1591 (LIBRARY).

[8] *Monumenta Pædagogica*, pp. 96-97.

[9] PACHTLER, *Monumenta Germaniæ Pædagogica*, vol. I, Berlin 1887, pp. 195-196: the regulation of the Collegio Romano of 1566 provides this teaching of elementary mathematics for the philosophers of the second year only, and sometimes by dispensation, for the "dialecticians" or philosophers of the first year.

[10] JOANNES REGIOMONTANUS, *De triangulis omnimodis* (SMITH, I, p. 260): the first work devoted exclusively to trigonometry (*c.* 1464).

[11] JORDANUS NEMORARIUS, appreciation of in SMITH, *Rara Arithmetica*, pp. 62-66.

Epytoma Joãnis
de mõte regio In
almagestũ ptolo
mei

Epitome of the Almagest of Ptolemy, 1564

Title page reproduced from the copy in the Library of the old Jesuits in Peking.

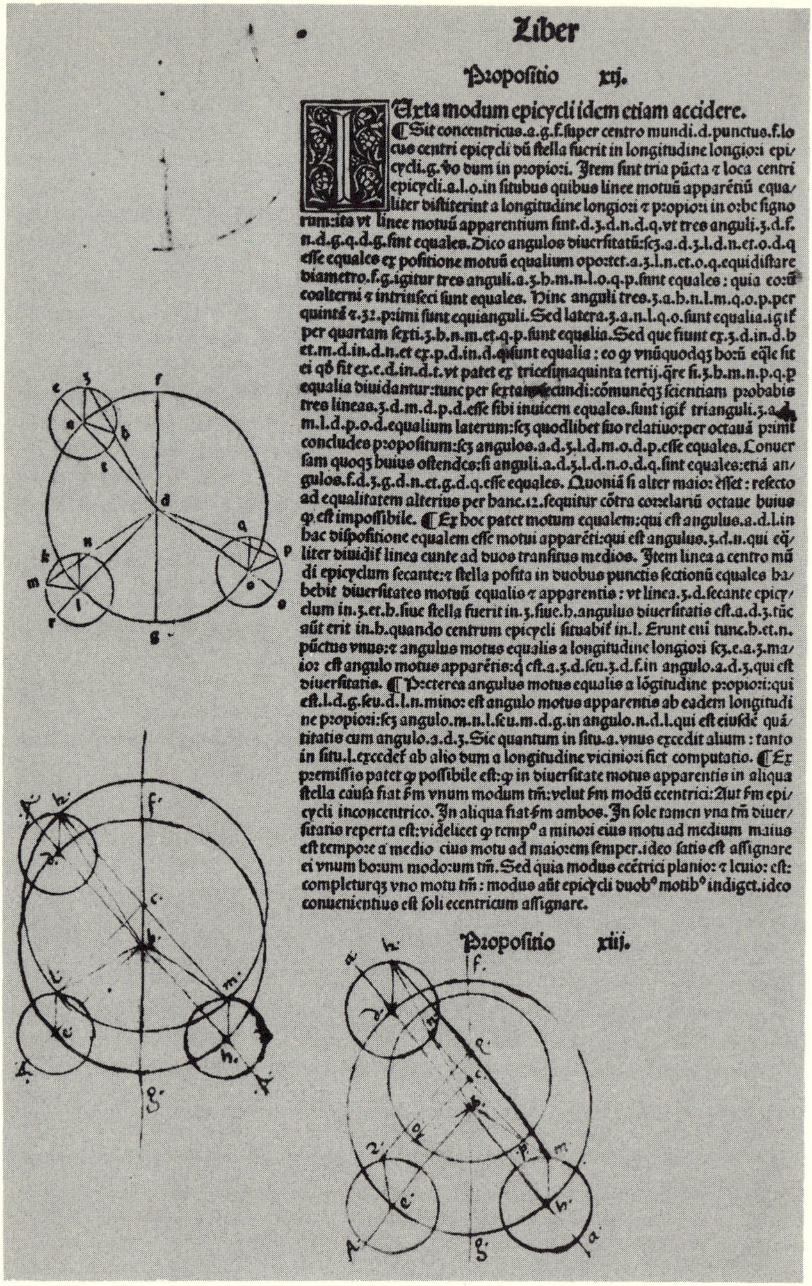

Liber

Propositio xij.

I Vxta modum epicycli idem etiam accidere.
¶ Sit concentricus.a.g.f.super centro mundi.d.punctus.f.lo
cus centri epicycli dū stella fuerit in longitudine longiori epi/
cycli.g.vo dum in propiori. Item sint tria pūcta ꝛ loca centri
epicycli.a.l.o.in situbus quibus linee motuū apparētiū equa/
liter distiterint a longitudine longiori ꝛ propiori in orbe signo
rum:ita vt linee motuū apparentium sint.d.ʒ.d.n.d.q.vt tres anguli.ʒ.d.f.
n.d.g.q.d.g.sint equales.Dico angulos diuersitatū:scʒ.a.d.ʒ.l.d.n.et.o.d.q
esse equales ex positione motuū equalium oportet.a.ʒ.l.n.et.o.q.equidistare
diametro.f.g.igitur tres anguli.a.ʒ.b.m.n.l.o.q.p.sint equales : quia corū
coalterni ꝛ intrinseci sunt equales. Hinc anguli tres.ʒ.a.b.n.l.m.q.o.p.per
quintā ꝛ.32.primi sunt equianguli.Sed latera.ʒ.a.n.l.q.o.sunt equalia.igiř
per quartam sexti.ʒ.b.n.m.et.q.p.sunt equalia.Sed que fiunt ex.ʒ.d.in.d.b
et.m.d.in.d.n.et ex.p.d.in.d.q.sunt equalia : eo ꝙ vnūquodqʒ horū eq̄le sit
ei qđ fit ex.e.d.in.d.t.vt patet ex tricesimaquinta tertij.q̄re si.ʒ.b.m.n.p.q.p
equalia diuidantur:tunc per sextam secundi:cōmunēqʒ scientiam probabis
tres lineas.ʒ.d.m.d.p.d.esse sibi inuicem equales.sunt igiř trianguli.ʒ.a.d.
m.l.d.p.o.d.equalium laterum:scʒ quodlibet suo relatiuo:per octauā primi
concludes propositum:scʒ angulos.a.d.ʒ.l.d.m.o.d.p.esse equales. Conuer
sam quoqʒ huius ostendes:si anguli.a.d.ʒ.l.d.n.o.d.q.sint equales:etiā an/
gulos.f.d.ʒ.g.d.n.et.g.d.q.esse equales. Quoniā si alter maior esset : resecto
ad equalitatem alterius per hanc.12.sequitur cōtra correlariū octaue huius
ꝙ est impossibile. ¶ Ex hoc patet motum equalem:qui est angulus.a.d.l.in
hac dispositione equalem esse motui apparēti:qui est angulus.ʒ.d.n.qui eq̄/
liter diuidiř linea eunte ad duos transitus medios. Item linea a centro mū
di epicyclum secante:ꝛ stella posita in duobus punctis sectionū equales ha/
bebit diuersitates motuū equalis ꝛ apparentis : vt linea.ʒ.d.secante epicy/
clum in.ʒ.et.b.siue stella fuerit in.ʒ.siue.b.angulus diuersitatis est.a.d.ʒ.tūc
aūt erit in.b.quando centrum epicycli situabiř in.l. Erunt eni tunc.b.et.n.
pūctus vnus:ꝛ angulus motus equalis a longitudine longiori scʒ.e.a.ʒ.ma/
ior est angulo motus apparētis:q est.a.ʒ.d.seu.ʒ.d.f.in angulo.a.d.ʒ.qui est
diuersitatis. ¶ Preterea angulus motus equalis a lōgitudine propiori:qui
est.l.d.g.seu.d.l.n.minor est angulo motus apparentis ab eadem longitudi
ne propiori:scʒ angulo.m.n.l.seu.m.d.g.in angulo.n.d.l.qui est eiusdē quā/
titatis cum angulo.a.d.ʒ.Sic quantum in situ.a.vnus excedit alium : tanto
in situ.l.excedeř ab alio dum a longitudine viciniori fiet computatio. ¶ Ex
premissis patet ꝙ possibile est:ꝙ in diuersitate motus apparentis in aliqua
stella causa fiat ſm vnum modum tm̄:velut ſm modū ecentrici:Aut ſm epi/
cycli inconcentrico. In aliqua fiat ſm ambos. In sole tamen vna tm̄ diuer/
sitatis reperta est:videlicet ꝙ temp⁹ a minori eius motu ad medium maius
est tempore a medio eius motu ad maiorem semper.ideo satis est assignare
ei vnum horum modorum tm̄.Sed quia modus ecētrici planior ꝛ leuior est:
completurqʒ vno motu tm̄ : modus aūt epicycli duob⁹ motib⁹ indiget.ideo
conuenientius est soli ecentricum assignare.

Propositio xiij.

A Page from the Epitome with Two Added Drawings by Ricci.

theory[12] or after Witelo,[13] the latter after Le Fèvre d'Etaples[14] or another author, if a more suitable one be found. In this lesson [of the second year] the applications of geometry and measures according to any author[15] may be omitted, etc.

3rd year.—Astronomy, beginning with [Purbach's] Theory of the Planets. Parts of the Almagest of Ptolemy[16] or at least of the Epitome of Regiomontanus,[17] the Alfonsine[18] Tables, the Astrolabe, etc., can always be added.

This document closes with a savoury observation which well characterizes the epoch: "The professor of mathematics will consecrate all his efforts on speculative mathematics; he is not to teach any astrology." In which one notices another survival of a past time: all the theorists of the Middle Age relied upon the authority of Boethius in defining music as a science (Acoustics in its infancy),[19] and the universities inscribed it after arithmetic, geometry and astronomy in the mathematical quadrivium.* Just as Boethius seemed formerly to felicitate Pythagoras for having undertaken a penetrating study of music, "relicto aurium judicio," neglecting the testimony of the ear, so Le Fèvre d'Etaples whom Nadal wished to take as a guide takes no account, in the art of music, of this special instinct which arithmetic could in no way supply!

Speculative music soon disappeared from the programmes of the Collegio Romano, but the essential part of Father Nadal's plan remained intact; the duration alone

---

[12] *Perspectiva communis* by GEORGES HARTMANN, Nürnberg 1542 (LIBRARY).

[13] On Witelo, cf. SMITH, II, p. 341.

[14] Le Fèvre d'Etaples (Jacobus Faber Stapulensis) was the last representative, and a somewhat notable one, of the decadent mathematicians of the Middle Age (SMITH, I, p. 307).

[15] These applications of geometry were specially developed in Italy (SMITH, I, pp. 294-306), for example in the rudimentary astronomical instruments (SMITH, II, pp. 344-368).

[16] *Almagestum Claudii Ptolomæi*, per LUCAM CAURICUM recognitum, 1528 (LIBRARY).

[17] Joannis de Monte Regio in *Almagestum Ptolomei*, Venice 1496 (LIBRARY).

[18] These *Alfonsine Tables* were reprinted in the *Prutenicæ tabulæ cælestium motuum*" of ERASMUS REINHOLD, Tübingen, 1551 and 1585 (LIBRARY).

[19] COMBARIEU, *Histoire de la musique*, Paris, 1924, t. I, pp. 221, 224.

* The old educational course of the Middle Ages divided itself into the elementary, or *Trivium*, consisting of Grammar, Rhetoric and Dialectic; and the more advanced *Quadrivium*, embracing Music, Arithmetic, Geometry and Astronomy.

of the studies varied. No doubt, St. Ignatius, in his Constitutions[20] had inscribed after Logic, Physics [Aristotelian], Moral Science, "also Mathematics" (etiam Mathematicæ), but all were not equally persuaded that these last acquirements suited the end of the Society of Jesus. We know this from the confidences of a Spanish Father, Gabriel Moralez, who, at the age of 36 years, taught mathematics at Vienna.[21] After having studied Greek, mathematics and medicine at Alcala, Paris, Louvain, Salamanca and Cordova, he had become a Jesuit about the month of July, 1559. The professor of the Collegio Romano was then Father D. Torres, who taught him the theory of the planets and some of the principles for the construction of sun-dials. Five months after his entry into the Society, Father Moralez himself took the professorship to teach "three books of Euclid, with the exception of the 4th and 5th propositions." In July, 1560, he was attached to the college at Vienna to explain there "the six books of Euclid [plane geometry], the manual of the Sphere, the use of the cosmographical globe[22] and of the astrolabe,[23] besides the ecclesiastical computus,[24] a treatise on the construction of sun-dials,[25] some questions in arithmetic and the first book of ordinary perspective."[26]

From that year, he says, at the rate of one lesson a day, up to the present I have reached the fifth book of Euclid, and I have explained a large part of the practical arithmetic of Gemma [Frisius][27] . . . I am too much occupied to make progress; I design nothing, I make no instruments, for want of leisure . . . . Besides, very few are fond of this study.

---

[20] *Constitutiones Societatis Jesu*, 4th Part, chap. 12, c.

[21] *Epistolæ P. Nadal*, II, 550-552 (in the collection *Monumenta Historica Societatis Jesu*).

[22] JOANNIS SCHONERI, *De usu globi terrestris*, 1551 (LIBRARY).

[23] JOANNES STOFFLERINUS, *Elucidatio fabricæ ususque Astrolabii*, Paris 1553 (LIBRARY).

[24] On the ecclesiastical computus, cf. SMITH, II, pp. 651, 665-669 in the LIBRARY: *Computus ecclesiasticus* of MAUROLICO.

[25] ANDREÆ SCHONERI, *Gnomonice*, Nürnberg, 1551 (LIBRARY).

[26] SMITH, II, p. 341, briefly explains this "ordinary perspective" of Peckham.

[27] This was the most popular arithmetic of the 16th century (SMITH, *Rara Arithmetica*, pp. 200-208).—In the LIBRARY, German edition of Frankfort, 1580.

Personally, I think that it is hardly suited to the institute of the Society, and as I find myself almost the only one occupied in this science in which truly in my opinion there is much that is unprofitable and useless, I was urged by the desire to apply myself to ordinary studies, leaving mathematics aside.

This not very flattering judgment was by no means that of the Superiors of the Society who took the firm stand that every Jesuit should acquire a minimum of mathematical knowledge. Father Polanco, successively secretary to three Generals, modified the time-table of the science classes for the young Philosophers of his Order :[28] at the end of the first year and at the rate of a lesson a day, two months for arithmetic; in the second year, four months Geometry, three months Sphere, two months Geography, three months Astrolabe; in the third year, four months Theory of Planets, three months Explanation of the Almanac [ecclesiastical computus], three months Perspective, two months Treatise on Sun-dials.

Father Polanco's project was applied by the professor of mathematics, Father Torres, with some slight modifications :[29] the share of arithmetic was increased by one month at the end of the first year; in the second year, they devoted " not less than four months " to the first three books of Euclid, four months to the Sphere and " three or four months " to Geography; the third year was divided into three equal parts of four months for the Theory of the Planets, the Astrolabe and Perspective.

About half an hour is appropriated to the repetition of the lesson, each working with compass and ruler to draw some figures; if four or five months remain over after the course, they study Books 4, 5, 6 and 11 of Euclid, review the Theory [of the Planets] and start the [Alfonsine] Tables; if there is time, they learn something about clocks, the [astronomical] year, the [geometrical] quadrant, the radius [or Jacob's staff], the ecclesiastical computation or a solid sphere.

This second regulation was not, it seems, the one followed by Father Ricci; at the Collegio Romano, in his

[28] *Monumenta Pædagogica*, p. 88.

[29] *Monumenta Pædagogica*, p. 477 : in the first year, for arithmetic, three months instead of two.

time, a third programme was adopted in which Arithmetic had been erased from the first year of philosophy :[30]

[In the second year of philosophy] : the first four Books of Euclid during four months approximately, Practical Arithmetic one month and a half, Sphere two months and a half, Geography two months, and, during the remainder of the year, Books 5 and 6 of Euclid.

[In the third year of philosophy] : Astrolabe two months, Theory of the Planets four months, Perspective three months, during the remainder of the time Clocks and the ecclesiastical Computation.

In these three successive regulations, some points should be noted. The course in Geography specially initiated the students into the mathematical part or, as it was then called, the "Cosmography" of the terrestrial globe, and it was principally based on the classical treatise of Ptolemy. However the ideas gathered by the Spanish and Portuguese explorers entered, thanks to printing, into general circulation :[31]

The cosmographers recorded them in their books and gave them a place on their maps. The small treatises of Peter Apian,[32] Vadianus, Gemma Frisius,[33] Glareanus[34] and others[35] . . . added one or more chapters on the "lands and islands recently discovered."

Modern maps were generally annexed, from 1508, to the editions of Ptolemy,[36] but it was only in 1570 that modern geography had its first special atlas : Ortelius was one of the first who conceived—one could not yet say, who realized—a radical reform of Ptolemaic geography . . . Alongside his name, is placed that of Mercator. . ; less erudite than Ortelius, he devoted himself more to the mathematical side of

[30] *Monumenta Pædagogica*, p. 478 : this would seem to be the programme to which Fathers Torres and Clavius conformed.

[31] VIVIEN DE ST. MARTIN, *Histoire de la Géographie*, I, 1873, p. 398.

[32] PETRI APIANI, *Cosmographicus liber*, republished by Gemma Frisius, Antwerp, 1529 (LIBRARY).

[33] GEMMA FRISIUS, *De principiis Astronomiæ et Cosmographiæ*, Antwerp, 1553 (LIBRARY).

[34] HENRICUS GLAREANUS, *De Geographia liber*, Cologne, 1581 and 1591 (LIBRARY).

[35] For example, JACOBI CHEYNERI ab ARNAGE, *De Geographia*, Douai, 1576 (LIBRARY).

[36] RUSCELLI's edition of Ptolemy, Venice, 1561 (R, I, pp. 469-471).

geography. . . . We owe to him also the system of projections by increasing latitude-lines employed almost exclusively in maps used in navigation.

Ricci later was able to use to great advantage what he had read or seen in Rome. He also learnt Perspective, that is to say the very special perspective system which had been built in the 15th century on the ruins of the geometrical science of the Greeks and which had been consecrated by the art of the Renaissance. The practice of drawing and painting sets man the problem of descriptive geometry : by means of two dimensions of a plane surface, to represent the three dimensions of space. In the monocular system of the Greeks, it is supposed that the picture is seen by one eye. Thence, by taking into consideration the three elements (the distance of the eye to the plane of the picture, the height of the eye relative to the objects to be represented and the distance or orientation of these objects in relation to the surface of the picture) one is enabled to draw on the picture, by means of geometrical constructions, the figure which reproduces the aspect of the objects represented.

From Perspective to Painting and Architecture is only a step : Ricci was an admiring witness of the efforts which the Papacy put forth to reform itself in the domain of art. The new art, often characterized by the style known as "baroque," placed Christian thought in the first rank of its preoccupations ; the expression of ideas there takes a preponderant place, relegating to a second level the simple representation of forms and the simple pursuit of plastic beauty. The Bolognese painters conceived a series of motives reproducing, in a very clear and very striking imagery, the principal scenes of the Old and the New Testament. Vignola created (1568) the Gesù which has remained to the present day the favourite type of the Italian church, the ideal model of the church unencumbered by any internal support ; it is like the crowning of the efforts carried on for many centuries. This art was reproached with want of originality : of the painters, they demanded lessons, and these soon became nothing more than professors; Vignola,[37] who is the great theorist of architecture, wrote its grammar, and his precepts too rigorously followed sometimes paralyzed in a certain measure the creative faculties of his successors. But this disadvantage was not displeasing to the new order of the Jesuits who did not preoccupy themselves so much in conquering minds through grace and charm as through the pulpit and the school.

The list of prescriptions of their programme of studies would be incomplete if we omitted one of the most original views of S. Ignatius:

> He had drawn the attention (of the Jesuits)[38] to the usefulness . . . of occasionally giving, at least in the universities, these courses with special stress on philosophy, mathematics and the other sciences which the Italian professors were especially partial to. These courses, which we may compare to our modern conferences, had less as their object to satisfy the necessities of the audiences as to establish the reputation of the house, provided that the greatest edification of one's neighbour and the service of God were always kept in view. He even foresaw the nomination of one or more special professors for this kind of demonstrations.

Very soon, the Congregations of the Holy Virgin and the Academies annexed to them associated their students with this form of scholarly exercises. Ricci was a member of them:

> The Academies[39] met every holiday in the morning. The first was called Parthenia, it treated of theological and philosophical questions, and had for its insignia Theology enthroned in a temple before which knelt Physics with a terrestrial globe and Mathematics with a celestial globe. Its motto was: "Leges impone subactis." The president of the Academy was nominated for three months, and, before relinquishing his post, discussed the subjects in public. Every Saturday, they recited the litanies of the Madonna in the church with the Ave Maria and other similar musical pieces.

[37] In the LIBRARY there are several editions of VIGNOLA, but of the 17th century.

[38] MISSON, *Les idées pédagogiques de St. Ignace de Loyola*, Paris 1932, p. 34.

[39] *Polanci Complementa*, II, pp. 87-88: letter of January 3, 1570 (in the collection of *Monumenta Historica Societatis Jesu*).

One Pope, Gregory XIII, Bolognese by birth (elected in 1572), incarnated the spirit of the time; a friend of the arts, a friend of the sciences, a friend of missions, he was able to surround himself with men most capable of realizing this ideal. Whereas a Palestrina for example devoted himself to reforming church music, Father Clavius applied himself to the correction of the calendar. Ricci never saw the achievement of these great enterprises, but he assisted in their genesis.

Having described the state of the sciences in Europe, and especially at the Collegio Romano, when Father Ricci was pursuing his studies there, it remains for us to portray the man who was pre-eminently his master, Father Clavius. Father Christophus Klau, better known by his latinized name of Clavius, was born at Bamberg in 1538 and, at the age of 17 years, was admitted by St. Ignacius to the Society of Jesus (January 1555); from Rome, he was sent to Portugal (October 1555) and there attracted attention by his scientific aptitude (in 1560, at Lisbon, he observed a total eclipse of the sun). Having returned to Rome, he completed his theological studies and at once devoted himself to teaching which in time to come was to bring him universal renown.[40]

Probably the man who did the most of all the German scholars of the 16th century to extend the knowledge of mathematics, although doing little to extend its boundaries . . . . . . He was an excellent teacher of mathematics, and his textbooks were highly esteemed because of their arrangement, particularly in the Latin schools. Sixtus V testified to his standing, saying: "Had the Jesuit order produced nothing else than this Clavius, on this account alone should it be praised." [41]

His pedagogic ideas form a valuable collection "on the manner in which mathematical teaching can be developed in the schools of the Company." [42]

---

[40] *Monumenta Ignatiana*, series IV, pp. 824, 836, 887.—SOMMERVOGEL, *Bibliothèque des écrivains de la Compagnie de Jésus*, II, pp. 1212-1223.—He assisted in the building of the country house, for the schoolmen of the Collegio Romano, near Santa Balbina.

[41] SMITH, I, pp. 334-336.

[42] *Monumenta Pædagogica*, p. 471.

It is necessary, first, to choose a master of exceptional erudition and reputation : . . and he should be invited to join in the public discussions and the more formal acts of promotion to the degree of Doctor. The students should understand that philosophy and the mathematical sciences are interrelated, as in fact they are : especially because, at the present time, they seem almost to despise those sciences, they do not pay any attention to them, even regarding them as useless.

Besides it is necessary that the master should have a liking for and an aptitude for teaching these sciences without being distracted by many occupations. . . And some gifted individuals should be chosen, who would be instructed, in a private academy, in various mathematical questions : otherwise, these studies far from developing will disappear, and moreover they bring great renown to the Society, since they very frequently recur in conversations and at the reunions of distinguished men. . .

Following the usage of the Society, the students are to apply themselves to these and at the same time to the Physica [of Aristotle], especially when the following questions are dealt with : number and movement of the celestial orbs . . . appearances of the stars following their conjunctions, oppositions and variations of distance, division to infinity of the continued quantity, flow and ebb of the sea, winds, comets, rainbow, halo and other meteorological phenomena, all subjects on which the calculators have written much.

The professors should know mathematics . . . It will be very useful if they avoid treating of questions which are of little assistance in understanding natural things, and which do great harm with their hearers to the esteem of mathematical teaching, for example : " mathematics have no demonstration, they make abstraction of the *ens* and the *bonum*."

In private conversations, the masters should exhort the students to apply themselves to these sciences, instead of diverting them from them, as the majority have done in preceding years. In this way, all the disagreements which strangers have noticed among ours will be caused to disappear.

It would be well once a month, for all the philosophers to meet together in one place and for one of them to pronounce a short eulogy of mathematical teaching. Then, with one or another, he would explain some geometrical or astronomical problem, or he might comment on some mathematical text from Aristotle or Plato, or he might bring forward some new demonstrations of certain problems of Euclid.

Toward the end of the year, an examination would be held.

There can be no doubt about it: Ricci was one of the faithful listeners who indemnified Father Clavius for the painful affronts to which the prejudice of certain Aristotelian "physicians" exposed him. Probably even, Ricci, not contenting himself with the usual measure, wished during his third year of philosophy[43] to be present at "the private lessons given in chambers, in order that he might become acquainted with the spherical elements of Theodosius,[44] of Menelaus[45] and of Maurolico,[46] with an introduction to the astronomical tables or the perpetual calendar, with a dial, ring or ray." Perhaps even, he was one of the assiduous who, in theology,[47] would pursue the study of the elements of Euclid "from the 7th to the 12th Books inclusive," and then add some of the conic sections of Apollonius of Perga.[48] Obliged to leave the Collegio Romano before he had finished his studies there, he seems to have been unable to initiate himself into the notions of Archimedes[49] and algebra[50] which Father Clavius reserved for the end of the theological course.

Ricci was gifted with a facile intellect and a lucid mind. He possessed moreover a very special gift of locality; this surprising mnemotechny seems to have been founded upon associations of visual images. He represented to himself fixed places, always the same and in the same order, for example the different apartments of a house he knew, and in these fixed places, he located in thought variable images (horse, lion, eagle, etc.); his mind, then running through in thought the places to which it was accustomed, found in them all the images he had placed there and in the same order. This ensemble of intellectual qualities enabled Ricci easily to assimilate the teaching of Father Clavius; he learnt how to handle the rule and the compass to solve graphically certain problems of geometric construction; probably he initiated himself into certain

---

[43] *Monumenta Pædagogica*, p. 478.
[44] THEODOSIUS: Rome edition 1591 (LIBRARY).
[45] In the works of MAUROLICO (LIBRARY).
[46] MAUROLICUS, *Opuscula mathematica*, Venice 1575 (LIBRARY).
[47] *Monumenta Pædagogica*, p. 474.
[48] Edition of FREDERIC COMMANDINUS, Bologna 1566 (LIBRARY).
[49] Basle edition, 1544 and (TARTAGLIA) Venice 1543 (LIBRARY).
[50] *Brevis regularum Algebræ descriptio, una cum demonstrationibus geometricis*, auctore JOANNE SCHEUTELIO, Basel 1550 (LIBRARY).

tricks in astronomy, horology, and practical constructions. That is all, and as has been said :[51] "It must be admitted that Ricci's science hardly ever went beyond the works of Clavius ;" we would be more categorical : "it usually remained well within them." He admitted it candidly later on : "Here [in China], they call me a prodigy of science, and that quite makes me laugh." [52] He wrote that he knew but little of astronomy, not enough to determine the course and the true situation of the planets, or to calculate eclipses and compose the calendar.[53] He conscientiously taught only ideas "easy and fluent for us," [54] and in particular he attributed to himself "little force and little knowledge." [55]

Ricci's modesty need not deceive us : his fellow students described him as "the mathematician," [56] but he was not what we now understand by that name. For him as for his master Father Clavius, scientific knowledge did not consist of specialities of intrinsic value ; they commended themselves especially as aids to higher studies.[57]

Mathematics do not seem only slightly suitable to the end which is set forth (in the Society of Jesus).

First of all, without Mathematics, our Academies would be deprived of a great ornament, still more they would be deficient for there is hardly any Academy of note in which mathematical exercises have not their place, and usually not the last place.

But above all, the other sciences have a very great need of it. Here is what they receive from it :

Poetry, the rising and setting of the stars.
History, the situation and the distances of various places.
Analytic Philosophy, examples of solid demonstrations.
Political Science, truly admirable methods of conducting affairs both private and during war.
Physics, the modes and diversities of celestial movements, light, colours, diaphanous bodies, sounds.
Metaphysics, the number of the spheres and of intelligences.
Theology, the principal parts of divine creation,
Law and Ecclesiastical Customary, precise determinations of time.

---

[51] BOSMANS, *Revue des questions scientifiques*, Louvain, January 1921, vol. 29, p. 143.
[52] R, II, p. 207. [53] R, II, p. 284. [54] R, I, p. 455. [55] R, II, p. 343.
[56] R, II, p. 395.
[57] *Ratio studiorum Societatis Jesu*, published in 1586 : in PACHTLER, *Monumenta Germaniæ Pædagogica*, 1887 : vol. II, pp. 141-143.

CHAPTER III

Ricci and Chinese Science

In the first periods of the 16th century people had been especially impatient to learn all about the new worlds that Christopher Columbus and Vasco da Gama had just revealed to Europe; the great navigators had seen more lands than they could give descriptions of, they had traversed seas and countries rather than studied them. What they had sought above all, was products which would enrich commerce; as well as points suitable for the establishment of colonies and the foundation of commercial factories. Secondarily and as if in passing, the mariner or the explorer had taken note of the physiognomy and the aspect of the countries visited.

But in the wake of the merchant[1] (and of the conqueror), there soon came the apostle of the Christian faith. Religious proselytism is the second dominating trait in the picture of the geographical history of the world in the 16th century. And we there find very great advance in the progress of our knowledge. It can never be loudly enough proclaimed how much the history of geography and the sciences owe to the completely disinterested labours of these devout and self-denying men. Appointed to live in the midst of peoples whom they wished to convert, sometimes constrained to adopt the costume and outward customs of the latter, obliged to render themselves masters of their language in order to preach the divine word, they had a thousand times more facilities than could the ordinary traveller to study at leisure and obtain a thorough knowledge of the countries and the peoples whereto their evangelical zeal had led them.

Among these diverse missionaries, shine in the first rank those who, to apostolic ardour, added a quite sound scientific training:

[1] Vivien de St. Martin, *Histoire de la Géographie*, pp. 401-402.

The first principle and the essential foundation of all Geography, said an Englishman of that epoch,[2] is *The History of Travel*, that is to say the reports written by travellers versed in Geography and Astronomy.

When the young Ricci disembarked at Macao (August 7, 1582), he seemed, more than any other, qualified to resume and develop the scientific essays, inaugurated by the Spanish cosmographers of the Augustinian Order.[3] In fact, since his departure from Rome (May 18, 1577), five years and more had elapsed. His experience had been considerably enlarged. In Portugal (from the month of June 1577 to March 24, 1578) he had been able to collect the traditions still quite fresh of the great cosmographer Pedro Nuñez.[4] During his sea voyages, he had begun the study of nautical science;[5] on all that he had observed, he had taken notes and drawn up memoranda of the route followed.[6] He had resided in the cities of Goa and of Cochin for more than three years and a half (September 13, 1578—April 15, 1582): and he had been requested to apply himself to the scientific study of this new and very imperfectly known country, in order to document his illustrious contemporary, Father Maffei, the "Titus Livius" of Portuguese explorations.[7]

The reputation of "mathematician" which Ricci had thus acquired among the missionaries[8] made him wanted in China, by Father Ruggieri; men in fact were required "capable of writing in Chinese ..... on certain mathematical questions of which that nation was very fond."[9] The young Italian missionary, only just landed at Macao (August 7, 1582), received orders to decipher an important collection of Chinese books with the help of

---

[2] TAYLOR, *Tudor Geography*, 1930, p. 41.
[3] BERNARD, *Missionnaires*, pp. 103 foll.
[4] GUIMARÆS, *Les mathématiques en Portugal*, 2nd edition, Coimbra 1909, pp. 16-21.
[5] TAYLOR, *A 16th century Manuscript Navigating Manual in the Society's Library* (*Geographical Journal*, vol. 78, 1931, pp. 346 foll).
[6] GILES, p. 367.
[7] BERNARD, *Goes*, pp. 17 foll.
[8] R, II, p. 395.
[9] R, II, p. 398.

the interpreters of the House of Catechumens,[10] and in the space of five months, he drew up for the Visitor, Father Valignano, a detailed report. He does not yet seem to have been acquainted except by hearsay with the state of the exact sciences. Like all Europeans who, having put into various maritime ports where intellectual life seemed to be absent, suddenly come upon the Chinese civilization, he was charmed and so to speak dazzled by the vision of that immense world which for a long time up till then had been forgotten by the West.

> Although their sciences, wrote Valignano,[11] are not so perfect as ours, they appear to have reached the degree achieved by the ancient philosophers, before Aristotle introduced into them method and before they had been clarified in the light of the Christian doctrine. . .

A little more than a year after his arrival at Macao, Ricci succeeded in establishing himself at Chao-ch'ing Fu, 肇慶府, the capital of the two provinces of Kuangtung and Kuangsi (September 10, 1583).[12] He there observed two eclipses of the moon (November 29, 1583 and May 24, 1584)[13] which had been "predicted very clearly and very exactly by another method than ours;"[14] and it was probably under the impression of this success that he wrote a truly excessive eulogy of scientific culture in China (September 13, 1584).[15]

> In their sciences, the Chinese are very learned: in medicine, moral physics [the science of morals], mathematics and astronomy, arithmetic, and finally all the liberal or mechanical arts. It is admirable that a nation, which has never had any relations with Europe, should have reached by its own means almost the same results as we with the collaboration of the whole universe.

He was not slow in correcting his too optimistic appreciations. His predecessor, Father Ruggieri, had been

---

[10] BERNARD, *Missionnaires*, pp. 206-207.—BERNARD, *Goes*, pp. 29 foll. "Le Ptolémée de la Chine."

[11] *Monumenta Xaveriana*, I, pp. 174-175.

[12] BERNARD, *Missionnaires*, pp. 195 foll.

[13] Observed by Ricci: R, II, p. 38.

[14] R, II, p. 45.

[15] R, II, p. 45.

able to captivate the benevolent curiosity of his learned friends by teaching them some of the more immediate practical applications of the European sciences, such as clocks, watches, and especially the glass-ware of Venice. In the eyes of people who knew nothing of Optics, the prisms of the common crystal passed "for precious stones of inestimable value or for a small piece of the sky," because one saw in them objects irradiated with all the colours of the rainbow.[16] Ricci did not scorn to "occupy himself with these trifles, leaving to others the care of great matters:"[17] a good clockmaker, a clever builder of houses, he was besides a correct draughtsman and expert mechanician.

Soon an opportunity presented itself for him to fathom the depth of Chinese learning. A small map of the world had been hung on a wall of his residence:

> Some scholars, wrote Ricci,[18] [after having seen it], asked me to draw a map of the various countries I had passed [in coming from Europe], to preserve a souvenir of my voyage. At that time, I was not completely familiar with the Chinese language; nevertheless, I selected some facts from the maps and the books I had brought with me, added notes and the journal of the route of several years and had the whole engraved on some blocks [of wood]. For the translation, I had been helped by an official interpreter; nevertheless, I did not succeed in avoiding mistakes altogether.

This Map of the World, inspired by European maps and retouched to suit the Chinese taste, set many problems for the curiosity of Ricci's visitors.

> They had not yet seen, wrote an old chronicler,[19] the earth distinguished (divided) by meridians and parallel degrees. They did not know of the equator, the tropics, the two poles and the five zones; and though they may have seen some of these circles on the sky or the celestial globe traced on the mathematical instruments, they had not yet seen them anywhere transferred to the earth. . . . .

[16] BERNARD, *Missionnaires*, p. 239.
[17] R, II, p. 60: October 20, 1585.
[18] R, I, p. 141.—BERNARD, *Missionnaires*, pp. 232 foll.—Giles, p. 367.
[19] RIQUEBOURG, p. 305.

In short, Ricci had to explain to them the "Geography" or more exactly the "Cosmography" after the fashion of the 16th century. He does not seem to have foreseen this case; he only possessed two books, the treatises on the Sphere of the world, composed by his master Father Clavius[20] and the Sienese Jesuit Alexandro Piccolomini;[21] he tried to recover from Chinese works everything that could help him to reveal to them the science of the universe.

The universe, he told them,[22] is constituted of celestial orbs and the elements, among which the earth stands for the foundation and the last heaven for the roof or cover.

Let us begin with the earth. It is very wide, since it measures several millions of Chinese *li*, which is equal to 6,300 of our leagues. The diameter which traverses it in the middle, is 2,000 leagues long and, in its centre, is the hell of the damned. . . . Since the earth is solid and spherical, it is surrounded by water like the yellow of an egg is by the white (in the shell). . .

As to the celestial orbs, without speaking of the three last the nearest of which to us contains all the stars, the seven others which are called orbs of the planets, are thus arranged:

First of all Saturn called in Chinese "earth star," 土 星.
2. Jupiter, "wood star," 木 星.
3. Mars, "fire star," 火 星.
4. the Sun
5. Venus, "gold star," 金 星.
6. Mercury, "water star," 水 星.
Lastly, the Moon.[23]

It must further be observed that time is measured by the movement of the sun and of the moon: in fact, the sun, by revolving around the earth, produces day and night. . . ; the moon, in approaching and receding from the sun by turns, sends back to the earth the light which it receives from the sun. When the moon is placed between the sun and the earth,

---

[20] *In Sphæram Joannis de Sacrobosco Commentarius*: Ricci probably did not possess the edition of 1581, but only that of 1570 or 1575 (SOMMERVOGEL, II, p. 1212).—That of 1585 is in the LIBRARY.

[21] *Sfera del Mondo*, frequently republished at the close of the 16th century: R, II, p. 72, note 3.

[22] R, II, pp. 510-511.

[23] CLAVIUS, *Sphæra*, 1585, p. 72.

it intercepts the rays and light of the sun; when, opposite to the sun, it passes into the shadow projected by the earth, it is suddenly eclipsed.

These astronomical truths—mixed as we know with so many errors—produced on the Chinese an effect analogous to that which caused the revelation of Ptolemy to the first savants of the European Renaissance: they opened infinite horizons on the grandeur of the Creation, the distribution of the worlds, the movements of the planets, the harmony of the spheres. The earth, round and immovable, is in the centre of the universe; around it are arranged the seven planets which revolve with unequal speeds; the firmament to which are attached the fixed stars and the signs of the zodiac, marks its visible limits. The orb of the stars and the orb of the planets have the form of a circle, because only the perfect form of the circle can express the perfection of the divine work. If one analyzes this universe, one finds in it only four elements: earth and fire, water and air; and, between the earth and the fire, intermediary elements allow of easy passage from the one to the other, since water which solidifies approaches the earth, and air which is heated approaches fire.

Ricci was able to materialize his teachings by constructing various instruments much in use in Europe: terrestrial globes,[24] celestial globes, astronomical spheres, sun-dials . . . On the terrestrial globes, the cartographical indications of the various regions are sober enough; the meridians, parallels and degrees, the equator, the tropics, the two poles and the five zones[25] are clearly distinguished on them. The celestial globe was probably constructed on the models which Ricci had formerly seen at the Collegio Romano.[26]

---

[24] R, II, p. 60: on October 20, 1586, he had constructed three globes.

[25] Riquebourg, p. 305.

[26] Stevenson, *Terrestrial and Celestial Globes*, 1921, vol. I, pp. 165-168: at the Collegio Romano there was a celestial globe attributed to Hieronymo de Boncompagni and a terrestrial globe by Emmanuele Filiberto (probably about 1570). Cf. two manuscript globes, the one terrestrial, the other celestial (1575), bearing the coat of arms of the Jesuits and two celestial globes by Cartaro (1577).

I do not know if I will succeed, he wrote on October 20, 1585,[27] since I can get help from no book: but whatever it may be, it seems to me that it will be superior to those of the Chinese. . . I have found among them twelve or, rather, twenty-four divisions of the year which correspond to ours in every respect: in these subjects, a little is worth a lot (*il puoco val molto*).

The spheres, constructed of brass and of iron,[28] are armillary spheres on which the earth occupies the centre; they are provided with two poles, the one fixed and the other movable.[29] The sun-dial, fixed on the façade of the residence, is vertical; it was an object of astonishment to the Chinese, who knew only of the horizontal dial.[30]

All these novelties intensely excited the curiosity of the scholars of Chao-ch'ing Fu, and the rumour began to spread that Ricci was the greatest savant in the world, "because, said an old historian, this people measures all the other nations by its own yardstick;"[31] the alchemists, stubborn with their chimeras, came to show him strange rocks or minerals,[32] merchants wished to learn how to manufacture velvet,[33] the common people regarded with a fearful reverence this magician whose clocks struck of themselves . . . Little by little, as time went on, a legend was created which made of him a tutelary genius of clocks in China.[34]

Truth to tell, as long as he resided at Chao-ch'ing Fu (September 10, 1583—August 15, 1589), Ricci did not so to speak meet any person capable of understanding the principles of European sciences and, if he taught the Christians the method of calculating their calendar,[35] it was by a quite empyrical method and without any demonstration.

---

27 R, II, p. 60.—R, II, p. 72: November 24, 1585.
28 R, I, p. 312.
29 CLAVIUS, *Sphæra*: frontispiece and page 24.—Cf. SMITH, I, p. 334.
30 R, II, p. 72.
31 RIQUEBOURG, p. 158.
32 R, II, p. 44.
33 R, II, pp. 65, 92.
34 PFISTER, pp. 29-30, note I.
35 R, I, p. 190.

For although it be commonly reported (outside this empire)[36] that many liberall sciences, and especially naturall and morall phylosophy are studies in China, and that they have Universities there, wherein such ingenuous artes are delivered and taught, yet, for the most part, this opinion is to be esteemed more popular than true: . . . . The people of China doe, above all things, professe the arte of literature; and learning it most diligently, they employ themselves a long time and the better part of their age therein . . . . . . their literature being . . . in maner infinite.

However, the moment was approaching when Ricci was in fact to discover a mind really disposed to learn the European sciences: this was a thoughtful and cultivated scholar, named Ch'ü T'ai-su, 瞿太素,[37] who prompted at first by the desire to study alchemy, placed himself under the training of Ricci for one year at Chao Chou, 詔州,[38] and afterwards remained in very close relations with him for two or three years[39] (between 1590 and 1595). The small library of European works had been enriched by at least two works of Father Clavius, his Practical Arithmetic[40] and his Elements of Euclid,[41] perhaps also his Treatise on Gnomonics[42] or Dials.

Ch'ü T'ai-su, relates Ricci,[43] wished to learn our sciences and especially mathematics. After our Arithmetic, which is much easier and methodical than theirs, he had me explain to him the Sphere of Father Clavius and the First Book of the Elements of Euclid. Finally, he learnt to make every kind of sun-dial, and also to measure heights and distances.

---

[36] HAKLUYT'S *Voyages*, EVERYMAN'S LIBRARY, 1926, vol. IV, p. 219: English translation of a Latin treatise on China, printed at Macao in 1590.

[37] This personage will receive prolonged attention in a work we are preparing on Matteo Ricci and the Chinese society of his time (*Matthieu Ricci et la société chinoise de son temps*).

[38] R, II, p. 102: in March 1592, Ch'ü T'ai-su had already left Chao Chou.

[39] R, II, p. 203.

[40] *Epitome arithmeticæ practicæ*, published in 1583, and 1585 (this last edition in the LIBRARY): SOMMERVOGEL, II, p. 1215.—SMITH, *Rara Arithmetica*, pp. 375-379.

[41] *Euclidis elementorum libri XV*, published in 1574, 1589, and 1591 (this last edition in the LIBRARY): SOMMERVOGEL, II, pp. 1213-1214.

[42] *Gnomonices libri VIII*, edition of 1581 (LIBRARY): SOMMERVOGEL, II, p. 1215.

[43] R, I, pp. 210-211.

At one stroke, these were new ideas, and especially absolutely new methods which appeared or re-appeared in China. In place of calculation by the abacus, "assuredly exact, but easily open to error and confined to but a few things," they could thenceforth count "with pen and paper:" Arabic numbers, four operations with whole numbers, the theory of fractions, problems in the Rule of Three, progressions, extraction of square root, . . . almost all that was to be found in the ancient Chinese treatises,[44] but had been completely forgotten by the scholars of the end of the 16th century. Ch'ü T'ai-su translated into elegant Chinese what Ricci had taught him in the common tongue: he thus created a scientific terminology which, in the course of time, eliminated almost completely the names employed by the ancient Chinese.

The First Book of the Elements of Euclid,—after a series of 36 definitions, 4 postulates and 20 axioms,—develops 48 propositions (34 theorems and 14 problems) on triangles, rigorously linked the one to the other, without any intermediary being neglected. Perhaps some high-flying mathematicians find this work very simple; but its simplicity constitutes its merit; it is within the reach of all, and China, after so many generations and so many peoples, proves in its turn that all can understand it.

By the help of these elements, Ch'ü T'ai-su was able to solve the elementary problems of the construction of dials or the measurement of lengths. Ricci further taught him to construct quite rudimentary instruments[45] which the astronomers and experts made use of, at an epoch when neither the telescope, nor the microscope, nor the sliding-gauge had yet been invented.

The most simple of all these apparatuses is the ordinary carpenter's square, sometimes called the geometric square. It is used to measure small distances by means of similar triangles.

The cross-bow, also called the geometric cross, golden rod, astronomical arm, Jacob's staff, . . . is much used

---

[44] WYLIE, pp. 168-169.
[45] SMITH, II, pp. 344-358.—BIGOURDAN, pp. 109 foll.

to estimate the height of stars at sea, in geodesy, and in topography. It is composed of a square shank or arrow with graduated marks, along which slides a perpendicular cross-piece.

The astrolabe is the best known of the ancient instruments; there are many models of it. To be strictly etymological, every instrument by which one can "take" the height of a star, is an astrolabe: thus the armillary sphere, formed of two and sometimes three circles whose planes are rectangular to each other, helps to determine the co-ordinates of the stars in relation to the equator and the meridian. The astrolabe recommended by Father Clavius is a "planisphere" that is to say essentially a disk (of stiff cardboard), graduated all round the circumference and provided with an open sight-rule or alidad turning about the centre in order to take sights of different points of the sky. What is most characteristic in the "plane-astrolabe" or "Planisphere" of Father Clavius, is what is called the "cobweb" or "network," a collection of curved lines obtained for a given latitude by stereographically projecting the celestial sphere on the plane of the equator. By this means are combined the advantages of an ordinary astrolabe and those of a celestial sphere.

The quadrant, closely related to the astrolabe, is reduced to a quarter of the circle; sometimes even, the arc is replaced by the two sides of a square on which the degrees are marked. As to what is called the "sextant," what we nowadays know by that name was only invented in 1730; in the 16th century the term was sometimes employed to designate a rotary compass, set at 60°.

Such are the instruments, truly very simple, of which we will now speak.

Ch'ü T'ai-su, says Ricci,[46] having transcribed the explanations in elegant Chinese showed them to his friends the magistrates, telling them that their books, in comparison with ours, were like people without brains. . . Everything enraptured him on account of its novelty. . He seemed quite

[46] R, I, pp. 210-211.—His works have not yet been recovered.

insatiable, passing day and night in study. In the books he published [in the course of time], he reproduced the designs of all that is contained in our works with their dimensions, and without error. Besides, he constructed many instruments: sextants, spheres, astrolabes, dials, compasses and other very fine things, in wood or in copper, many even in silver.

This apprenticeship was not a matter of a few lessons, but Ricci never regretted having devoted a considerable time to it because, wherever he went afterwards, he found the way prepared by Ch'ü T'ai-su who constituted himself "the herald and forerunner" of his master. In the month of April 1595 (nearly thirteen years after his arrival in China), the Italian missionary yielded to the entreaties of his friends who urged him to leave the province of Kuang-tung, "rough and but little civilized," and he went and established himself in the city of Nanch'ang, 南昌, the capital of Kiangsi (June 28, 1595). His renown drew him almost immediately a large number of disciples eager to receive instruction in his school: some to study alchemy, others to learn the secret of mnemotechny, and several to study our sciences.

It was still a recent work of Father Clavius' which seems to have inspired his teaching.[47] Ricci had noticed that his hearers knew only the "equatorial dial,"[48] "for the 36° of latitude which was the same throughout the world." He constructed a "horizontal dial"[49] (for the latitude of Nanch'ang 29°) on which are indicated "the signs of the zodiac, the length of the days and nights, the hour of sunrise, the moment when the sun enters a celestial constellation and reaches the zenith." He then explained that the height of the pole is not the same for all places on the terrestrial globe and constructed a second horizontal dial for the town of Ningpo in 32° North latitude,[50] and later, a third for the 31° of latitude.[51]

[47] *Fabrica et usus instrumenti ad horologiorum descriptionem peropportumni*, published in 1586: Sommervogel, II, p. 1216.—Cf. R, II, p. 217, note I.
[48] Clavius, *Gnomonices*, 1581, pp. 273-293.—Cf. R, I, p. 17.
[49] *Ib.*, 142-204, 575-584.—In August-December 1595: R, I, pp. 264-265; II, pp. 184-185.
[50] November 4, 1595: R, I, p. 269; II, p. 208.
[51] October 12, 1596: R, II, p. 216.—R, II, p. 224.

Ricci's scientific reputation increased daily: "In truth, he wrote on October 28, 1595,[52] I may say that, for them, I am another Ptolemy," and he enumerated their "absurdities," according to the ideas current in Europe at the end of the 16th century.[53]

1.—The earth is flat and square, the sky is a round canopy: they did not succeed in conceiving the possibilities of the antipodes.

2.—There is only one sky [and not ten skies]; it is empty [and not solid], the stars move in the void [instead of being attached to the firmament].

3.—As they do not know what the air is, where we say that there is air, they affirm that there is a void. By adding metal and wood, and subtracting air, they count five elements [and not four]: metal, wood, fire, water and earth. Still worse, they make out that these elements are engendered the one by the other, and it may be imagined with how little foundation they teach it, but as it is a doctrine handed down from their ancient sages, no one dares to attack it.

4.—For eclipses of the sun, they give a very good reason: namely that the moon, when it is near the sun, intensifies its light! On the contrary, when it is diametrically opposed to it, it is so terrified that it loses its light and its colour! Other scholars say that there is an empty hole in the middle of the sun and, when the moon is opposite that hole, it cannot be illuminated!

5.—During the night, the sun hides under a mountain which is situated near the earth.

Ricci revived then, for the use of the Chinese, the teachings which had been current in Europe for nearly two thousand years. Plato had set the mathematicians this problem: "What are the circular and perfectly regular movements which should be taken for hypotheses in order to preserve the appearances presented by the wandering stars?"[54] Now, since Plato placed the Earth in the centre of the celestial circulations, and supposed it

[52] R, II, p. 175.
[53] R, II, pp. 175, 184-185, 207.—R, I, pp. 311-312.
[54] Duhem, *Histoire des doctrines cosmologiques de Platon à Copernic*, vol. I *passim*.

to be stationary there, the problem could still be formulated thus: "Place several spherical concentric globes one inside the other; give to each of them a uniform rotation round a suitably chosen axis; suppose that the movement of the inside orb combines with the movements of those surrounding it; then combine these movements in such a way that the resulting progress of a star fixed to the orb nearest to the centre represents the apparent movement of the observed planet." It was under this form that this problem gave birth to diverse systems of homocentric spheres. That is the problem of astronomy as it was understood till Kepler. For us who know that for the circle, must be substituted the ellipse, it is evident that the cosmological doctrine built on this basis by Aristotle did not solve, and could not solve, in spite of its grand unity, the problem set by Plato; it did not preserve, it was doomed not to preserve, the totality of the celestial appearances. To explain more minutely the apparent movements, Greek Antiquity, the Islam middle age, the Christian middle age and the beginning of the Renaissance had seen the development of the system of eccentrics and epicycles, persisting in supposing that the circle is the noblest of geometrical lines, and trying to follow dumbly by combinations of circles the real movements of the stars.

Ricci dared not teach the theory of eccentrics and epicycles to his hearers at Nanch'ang, but he strove to make them understand the essential points of the cosmological problem. For their use, he reproduced the schema of the nine superposed spheres, he translated the theory of the four elements . . .[55] Imbued as he was with the Greek method, essentially based on the ecliptic, he noticed very soon that the Chinese method is founded especially on the equator, and, on the hypothesis that the earth is stationary, he explained to them "the two poles, one mobile, the other immobile."[56]

Since the Zodiac [or Ecliptic] is oblique in relation to the Equator, the pole of the Zodiac does not coincide with the pole of the Earth.[57] Then, since the eighth sphere moves

[55] R, II, p. 265. [56] RIQUEBOURG, p. 305.
[57] CLAVIUS, *Sphæra*, 1585, p. 303.

round the axis of the world, the Zodiac which is one part of the eighth sphere, will circumscribe the pole of the world. The circle described by the pole of the Zodiac round the North Pole is called the Arctic (Celestial) Circle; and the circle described by the other pole of the Zodiac round the South Pole, is called the Antarctic (Celestial) Circle.

This distinction, which clarifies to a great extent the enigmas of Chinese astronomy, was willingly admitted by his hearers. Other points brought forth objections: it seemed to them incredible that the sun could be greater than the bottom of a cask,[58] but they ended by being convinced at least that it is of greater dimensions than the earth, "for certain scholars, having measured it with their instruments, have attributed to it more than a thousand leagues."[59] The stars, adds Ricci, are still greater, many greater than the earth: "that seems to them an incredible paradox as the fire situated under the heaven occupies the highest part of the elementary earth, where it gives birth to the comets and the shooting stars!"[60]

And in truth, he says,[61] if China was the entire world, I would not hesitate to call myself the greatest mathematician and even the greatest philosopher of nature, because all that they say is ridiculous and I am surprised at the little they know: they are all addicted to moral philosophy, and they have cultivated elegance of discourse, or, rather, of style. . . They are persuaded that I am a phenomenon of scientific knowledge and that no other like me has ever come from our country: which causes me great amusement.

Ch'ü T'ai-su continued, by his letters and writings, to fan the fire which he had kindled.

Seven years ago, he wrote to Ricci on May 19, 1596,[62] in passing through Nanch'ang, I extolled your rare merits to all the scholars of the place. . . One of my disciples, called Helo (?) . . . . . wished to renounce the passing of the examina-

---

[58] R, II, p. 207.
[59] R, I, pp. 311-312.
[60] RIQUEBOURG, pp. 306-307.
[61] R, II, p. 207.
[62] Inserted in the long letter of Father Longobardi, October 18, 1598: for ex. HAY, *De rebus japonicis, indicis et peruanis epistolæ recentiores*, Antwerp 1605, pp. 918-919, gives a Latin translation from the Italian; it is difficult to find the Chinese characters corresponding to the romanizations of the proper names.

tions for the licentiate in order that he might attach himself to you. . . I dissuaded him from doing so because his mother was old. . . He successfully passed the examination and is now the fourth Governor of Caucea (?) . . , in the province of Kuang-tung. Perhaps he has paid you a visit ? . . . Last year I met an Academician to whom you had explained the [Euclid's] definitions of the point, the line, the extremity [of a line], thickness. He is full of admiration for it. . . Many scholars [of Soochow in Kiangsu, the residence of Ch'ü T'ai-su] wish that you would go there. . .

This year, having enough leisure, I gathered together and put in order what you had taught me, and of it compiled a book which I showed to the Academy of Scholars: there was no one who did not admire your doctrine. . . I send you the manuscript for you to correct it. . . and afterwards I shall publish it. . . ; and thus our friendship will not be without some result. Here, they esteem especially the books of Hothu (?), Coscui (?), Pequa (?), Queusciu (?), Thaiquitu (?),* and others of the same kind which discuss the point, the line, the extremity and thickness. All the scholars affirm that the line [the circumference] is the circle, but, according to your teaching, the line [the circumference] is that which bounds the circle, and the circle is included in the inside [of the line].

The eclipse of the sun of September 22, 1596, completely established the astronomical reputation of Ricci at Nanch'ang.[63]

Several months before, it had been announced by the Board of Mathematics at Peking. On that occasion, the magistrates had to assemble in order to bring help, so they said, to the sun and the moon, by striking bronze instruments until the eclipse ended. Now it was much less than had been foreseen. Many people came to ask me the reason : had there been an error in the calculation, or had its importance been reduced through other causes ? On that occasion, I explained to them that the eclipse of the sun is not universal and that it can be greater in one place than in another ; thus it might happen that it would be greater in the place [Peking] where the calculation had been made whereas it was smaller in this province [Kiangsu] which is much more central. They were

---

[63] R, II, p. 224—Souciet, II, pp. 183-204 reproduces the calculations of that eclipse made by Hsing Yün-lu, 邢雲路, following the *Ta T'ung-li* of Kuo Shou-ching, 郭守敬.

* Evidently : *Ho T'u* 河圖, *Lo Shui* 洛水, *Pa Kua* 八卦, *T'ai-chi T'u* 太極圖, *Chiu Ch'ou* 九疇.— *Translator's Note.*

contented with this explanation because it did not contradict the King's mathematicians and the reasons that I gave them for eclipses of the sun and the moon satisfied them: all that had been unknown to them, especially the cause of the eclipse of the moon which their scholars could never explain.

After the month of October, 1596, a new book by Father Clavius on the Astrolabe[64] which Father Valignano had had bound in India was received. The author's preface extols the merits of this instrument:

Of all the instruments which one uses . . for the prime mover [the Sun], none appears to me to surpass that named by Ptolemy the Planisphere, commonly called the Astrolabe. There in fact, all the celestial circles of the prime mover are geometrically projected on the plane in such a manner that each of their points and of their arcs can be measured with as much care and precision as on a perfectly spherical globe. Without doubt, the solid sphere or globe . . . is the most perfect image of the whole heaven . . . but it is difficult to construct and keep perfectly round. . that is why, a plane instrument is much more convenient for travellers, because it can be easily transported from one place to another and preserved without injury. Besides, all the necessary circles can be drawn on it.

In 1597, Ricci made haste to use this book[65] not only to show it to many of the scholars who were quite surprised at the accuracy of the figures, but also in order to construct another dial. . . on which were indicated the duration of the day and night for each of the twenty-four signs [of the Zodiac] and their half-intervals; moreover, the time of the rising and setting of the sun, and, which most of all astonished the scholars, the twilights.[66]

After so many years' experience, the missionary concludes:[67]

The Chinese do not possess any sciences: one may say, only mathematics are cultivated and the little they know of that is without foundation, they borrowed it from the Sarracens; only the King's mathematicians teach the science to their sons. They just manage to predict eclipses, and even in that make many mistakes.

---

[64] *Astrolabium*, edition of 1593 (there are three copies of it in the LIBRARY; one of them bears this inscription: " P. Mathæo Riccio donum auctori alla China " and numerous corrections)—R, II, pp. 217 and 241: on October 12, 1596, Ricci had not yet received this book; on December 25, 1596, he had already been using it for some time.

[65] R, II, pp. 241-242.

[66] CLAVIUS, *Astrolabium*, pp. 657-663.

[67] R, II, p. 237.

CHAPTER IV

THE PROBLEM OF CHINESE ASTRONOMY

SINCE astronomy became a mathematical science, with no direct use for the individual, banished from tuition and reserved for specialists, a false idea was formed as to its origins.[1] To observe the movement of the stars was formerly a concrete and very simple operation : in primitive civilizations, the astronomy of celestial appearances is the primordial science whereas the physico-chemical sciences operating in a domain in which the impressions of the senses are constantly bewildered, are almost inaccessible to uncultivated minds. It is therefore not surprising that Chinese astronomy attained from the remotest antiquity a considerable development whilst physics was summed up in the antithesis of the two principles, the *yang* 陽 (warm, dry, male element) and the *yin* 陰 (cold, humid, female element), and that chemistry contented itself with the evolution of the five elements (earth, wood, metal, fire, water).

On the other hand, the ancient astronomy is a primordial element of Chinese civilization, for all the general ideas are crystallized in the shape furnished by contemplation of the sky. It is therefore difficult to treat of it, as one would speak of a purely objective Western science, without mentioning, at the very least, its intimate connexion with the philosophical beliefs.

The history of Chinese astronomy is composed, as is the general history of China, of two periods, the ancient and the modern, separated by the Burning of the Books (213 B.C. by order of the founder of the Ch'in dynasty).

In the ancient era, astronomy had especially a metaphysical, political, ritual and social importance, making of the ruler the

[1] According to the various articles by M. DE SAUSSURE.

Vicar of Heaven on earth, entrusted with maintaining here below " the order and regularity of celestial things," notably the promulgating of the calendar.

This character of the ancient astronomy profoundly marked the modern astronomy and it has been noted by all observers. Ricci for example never ceases saying that " astrology " was generally practised by the Chinese society of his time, and it would have been an error not to see in this somewhat inappropriate term all the social importance and philosophical elevation with which it was clothed in the Far East.

It is not the sun, it is the sky, it is the firmament which produces the seasons and the changes of vegetation; this physical influence extends to the domain of morals. Thus the terrestrial sovereign, the Son of Heaven, can create trouble in the celestial movements, if he deviates from his duties. In consequence of the same idea, the terrestrial sovereign, the Vicar of Heaven, presides alike over physical laws and moral laws on the earth: when a predicted eclipse does not take place, the Emperor is congratulated on it!

This point must be well understood in order to appreciate the religious importance of the drawing-up of the calendar, the privilege of the Son of Heaven. The imperial domain being on earth the correlative of the Pole Star in the sky. . things of the Heaven and the Earth are indissolubly answerable to each other. Thus the promulgation of the calendar is not envisaged from the single utilitarian point of view of the computation of epochs, but especially from the religious point of view of the harmony between the Heaven and the Earth. The first duty of the Sovereign is to make known to the people the conditions of this harmony, that is to say to indicate the limit of the seasons and the situation of the terrestrial (civil) year in relation to the celestial (astronomical) year. If he fails in this duty, if he neglects the calendar, not only will there result inconvenience to the people; he commits in the sight of Heaven, whose vicar he is, an irreverence which endangers his divine right, his celestial Mandate and which to that extent diminishes his VIRTUE that is to say the virtuous power of his dynasty.

But the Emperor, although Son of Heaven, is but a man like other men; he is by no means regarded as omniscient. To govern well he must before everything choose his ministers well. If his ministers counsel him ill, he is responsible to Heaven for their errors, but they are responsible to him. The drawing-up of the calendar being the first duty of the sovereign to Heaven and his people, if errors occur through the negligence of his ministers or the astronomers, the Emperor is responsible for them to Heaven, but those ministers or astronomers are responsible to him; for they have committed a fault the consequences of which to the dynasty will be very serious.

Thus, at all times, we see that the most effective political pretext which an emperor could use to crush a too-powerful minister is to accuse him of having neglected the calendar; for he is sure of finding in public opinion a moral support based on the religious sentiment of the importance of the committed fault. In analogous cases and for the same motive, our ancient kings did not fail to bring an action for sorcery.

Whereas with the Greeks, the astronomer is a philosopher, a "friend of the truth" (Ptolemy's expression in speaking of Hipparchus), a scientific individuality without official mandate, most frequently slightly at variance with the city clergy; in China, on the contrary, astronomy is intimately allied to the sovereign pontificate of the Son of Heaven. It is a function of the State, it is the expression of social and religious order. And that conception, without being formulated in as explicit a manner in modern times as in the ancient writings, profoundly penetrates both public opinion and popular sentiment.

The Gregorian reform of the Julian calendar had been effected in Europe chiefly by Father Clavius and it had been promulgated at Goa on November 14, 1583. Ricci availed himself of it[2] for the private use of the Christians. In 1592 or 1593, at Chao Chou, the President of the Board of Rites at Nanking, passing through that city on his way to the island of Hainan, expressed the desire to take him with him when he returned to the capital:[3]

> to be helped in correcting their year or calendar which was often erroneous, and thereby to acquire a very great reputation.

This invitation had no practical effect until after several years (June 1598),[4] but it was from that moment,

---

[2] R, I, p. 190.
[3] R, I, pp. 234-235.
[4] R, I, p. 281.

it would seem, that Ricci understood completely how much this means of access might serve to penetrate further into Chinese society and to make himself in some measure a naturalized Chinese. Here, in fact, is what he wrote on May 12, 1605[5] to a correspondent in Europe:

I address to Your Reverence urgent prayers for a thing which I have for long requested and to which I have never received any reply: it is to send from Europe . . . a Father or even a Brother who is a good astronomer. I say, "astronomer," because as regards what concerns the other subjects of geometry, sun-dials and astrolabes, I know enough for my use. . . but these questions are held in less esteem than the trajectory and the true situation of the planets, the calculation of eclipses and, on the whole, the art of constructing ephemerides. This is explained by the fact that the King maintains, I believe, more than two hundred people at great expense, to calculate the ephemerides each year. . . . These people know nothing more than to lay out their numbers in accordance with their rules; they do not understand the meaning and, if the result is erroneous, they content themselves with saying that they have conformed to the precepts of the ancients.

With our maps of the world, dials, spheres, astrolabes and other instruments which I have constructed and explained, I succeeded in acquiring the reputation of being the greatest mathematician in the world. . . Consequently, if this mathematician were to come to China, after we had translated our tables into Chinese, which would be very easy for me, we would undertake the task of correcting the calendar, and, thanks to that, our reputation would go on increasing, our entry into China would be facilitated, our sojourn there would be more assured and we would enjoy greater liberty.

If one takes into account that at that time it required five, six or even seven years to receive in China a reply to a letter sent to Europe, one would hardly be mistaken in dating back to the year 1598 the first request for an associate astronomer. At that time, after a furtive sojourn in Peking (September 7—November 5), Ricci returned to the Yangtzŭ River region and, on February 6, 1599, under the protection of Ch'ü T'ai-su, peacefully re-entered the

[5] R, II, pp. 284-285.

city of Nanking, the former capital. He had been preceded there by an extraordinary fame and all wished to admire the presents which he intended to get offered to the Emperor: the time-pieces or "clocks which strike of themselves," the oil paintings, the manichord or spinet, the triangular crystal prisms and other curious objects.

Without delay, he recommenced his teachings at Nanch'ang Fu, refuting the "thousand chimeras" which the scholars used to explain eclipses,[6] and he gave such clear and manifest reasons that a large number of people declared themselves convinced.

The height of the pole differs according to the climates, and, similarly, the length of the day and night, the time of the rising and setting of the sun. . . Thanks to the theory of the two poles, the one movable, the other fixed, they received great help in understanding the movement of the planets. . One can construct horizontal dials, and others fixed on no matter what wall, and if one draws on them the celestial signs or the twenty-four divisions of the year with its parabolas, hyperboles and straight lines, the shadow of the style would not diverge an atom for the day indicated. . . They are also shown geometrical quadrants and other means of measuring quite easily the height of towers, the depth of wells and valleys, the extent of distances.[7]

It was perhaps at this time that some friends printed for the first time the small treatise in which Ricci combats the error of the five elements and demonstrates that not more than four of them could be elements.[8]

His success was considerable; three pupils joined Ch'ü T'ai-su to take methodical lessons in the exact sciences.[9]

The two first, disciples of a great scholar, a native of Kiangsi, named Li Hsin-chai, 李心齋, were well versed in Chinese mathematics, on which one of them had even written a book, published under the assumed name of the son of Li Hsin-chai.

---

[6] R, I, pp. 311-2.

[7] Clavius, *Astrolabium*, 1593, pp. 68, 74-75, 88-90, 125, 128-129, 133-135, 151-154.

[8] Clavius, *Sphæra*, 1585, 33-38.—Cf. Pfister, 40, No. 20.

[9] R, I, pp. 313-316.

The third, (Cian Ianmue ?), still more intelligent than the two preceding had been sent to the Father by an Academician (Guansciunngan ?) who, in the retreat where he lived some three or four days' journey from Nanking, occupied his leisure in trying to introduce a little method into Chinese mathematics. . . .
To this very small group of pupils, Ricci began by devoting the time left him from his outside engagements. The one who profited most from his teaching was the third, of an intractable enough nature, but who quickly felt himself dominated by the Father and did not dare contradict him further. . . By himself, he succeeded in understanding the First Book of Euclid in Ch'ü T'ai-su's translation, and wished no longer to hear any arguments except after the method of Euclid ; Ricci did not have the time to explain more of it to him, he had to content himself with learning all that he could ; then he returned to his master who, without delay, published some very important books in which he reproduced many of the mathematical and moral teachings of the Father.
At the same time, with the help of his pupils, Ricci constructed some plane-dials and others of different forms, as well as many spheres, terrestrial globes, geometrical quadrants . . . . and sextants. . . .

During the fifteen and a-half months that he resided at Nanking (February 6, 1599—May 18, 1600), he did not cease discreetly investigating the problem of the reform of the calendar. Three categories of adversaries were able to oppose him : the bonzes, the Emperor's mathematicians and the scholars ; he cleverly disarmed them.

The bonzes, wishing to explain the periodic return of day and night, affirm that the sun hides itself behind Mount Sumeru which is situated in the sea 24,000 miles under the water. Besides they divide the earth into four parts, and place China in the southern region. Then they claim to give an account of the eclipses of the sun and the moon in this manner : a divinity, named Holohan, produces the eclipse of the sun by covering it with his right hand, and that of the moon by covering it with his left hand ! [10]

Cian Ianmue advised the Father not to refute the bonzes directly, but to confine himself to teaching

[10] HERMANN, *Buddhistische Karten.* (SVEN HEDIN, *Southern Thibet*, VIII, 1922, p. 282.)

mathematics: "thus, he said, one could easily convince one's self that those who so grossly deceived themselves in what concerns natural phenomena and the things of this life, could not be believed when they spoke of the supernatural things of the other world."

Not only in Peking, he wrote,[11] but in this capital also (Nanking) there is a College of Chinese Mathematicians, and this one certainly is more distinguished by the vastness of its buildings than by the skill of its professors. They have little talent and less learning, and do nothing beyond the preparation of the almanacs on the rules of calculation made by the ancients; and when it chances that events do not agree with their calculation they assert that what they had calculated was the regular course of things, but that the aberrant conduct of the stars was a prognostic from Heaven of something going to happen on earth. This something they made out according to their fancy, and so spread a veil over their own blunders. These gentlemen did not much trust Father Matteo, fearing, no doubt, lest he should put them to shame; but when at last they were freed from this apprehension they came and amicably visited the Father in the hope of learning something from him. And when he went to return their visit he saw something that really was new and beyond his expectation.

There is a high hill at one side of the city, but still within the walls. On the top of the hill there is an ample terrace, capitally adapted for astronomical observation, and surrounded by magnificent buildings which form the residence of the Professors. . . . On this terrace are to be seen astronomical instruments of cast-metal, well worthy of inspection whether for size or for beauty; *and we certainly had never seen or read of anything in Europe like them.* For nearly 250 years they have stood thus exposed to the rain, the snow and all other atmospheric inclemencies, and yet they have lost absolutely nothing of their original lustre. . . .

Some years afterwards Father Matteo saw similar instruments at Peking,[12] or rather the same instruments, so exactly alike were they, insomuch that they had unquestionably been made by the same artists. And indeed it is known that they

[11] R, I, pp. 315-317. TRIGAULT, *De Christiana Expeditione apud Sinas. . . .* Lyons, 1616. Quoted by Col. Yule in *Travels of Marco Polo*, Third edition, 1929, Vol. 1, p. 451.
[12] R, I, p. 24.

were cast at the period when the Tartars were dominant in China; and we may without rashness conjecture that they were the work of some foreigner acquainted with our studies.

We said above that these instruments were doubtless constructed by Kuo Shou-ching in his school of young Mongols. Towards 1279, Kublai ordered three copies of them to be cast for his residence at K'ai-fêng Fu, and later, the founder of the Ming, when rebuilding his capital Nanking,[13] probably transported there the copies that Ricci admired.[14]

The notoriety of the missionary grew daily, and often unknown to him. The Chinese Map of the World printed at Chao-ch'ing Fu had gradually been disseminated throughout the whole of China and the Viceroy at Nanking, secretly, had had an impression on stone made at the city of Soochow;[15] at Nanch'ang, the Father had himself republished this first work introducing into it certain improvements, but that was not enough to satisfy his friends at Nanking who impatiently desired new annotations and a larger size. A high mandarin, "Uzohai," [Wu Tso-hai, 吳左海] adorned this reissue with a eulogistic preface.

> I have ascertained that the maps which he (Ricci)[16] makes are all based on originals printed in his own country. . . . . Our philosopher is certainly no devotee of the black art or superstitious practices; he worships God morning and evening, and has sworn never to be false in thought, deed, or word. With regard to his mathematical demonstration of the nature of the heavens and the remoteness of the sun, moon, and stars, although it is not easy to follow, yet there seems to be evidence for his statements, and they should be recorded until fuller knowledge is available.

On January 24, 1601, Ricci entered Peking for the second time; in his baggage,[17]

> were many books on mathematics which he had put together to use in case the King wished to attach him to the

[13] GAILLARD, *Nankin*, pp. 218-219.
[14] Cf. supra page 5.
[15] R, I, p. 286.
[16] GILES, p. 376.
[17] R, I, pp. 364-365.

Board of Mathematics at Peking. In China it is forbidden under pain of death to study mathematics without the King's authorization, but this law is no longer observed: that is why the eunuch Ma T'ang . . . had them placed in a box and sent to the guardroom of the Treasurer in the citadel at Tientsin with this superscription: "Books which the eunuch Ma T'ang, supervisor of the royal custom-house, found in the baggage of a foreigner, Matteo Ricci: these books are forbidden by the laws of the kingdom because they treat of mathematics."

Through a providential mistake, the servant of the governor who was beseeched by Ricci to restore the baggage could not read and he had returned this precious collection of scientific books when the imperial order came to conduct him to Peking.[18] The founder of the mission to China was more than ever persuaded of the necessity of this indirect form of apologetics, and, in spite of the opposition he met with from some of his most faithful collaborators, he never ceased, during more than nine years, to endeavour to ingratiate himself sufficiently with the Emperor in order that he himself or another Jesuits should be authorized to reform the calendar and the ephemerides. The scholars who aided him in this were all—or nearly all—natives of Kiangsu and Chêkiang.

The first of the important persons with whom he entered into relations on this subject, was Fêng Yingching,[19] who, an official in the province of Hukuang in 1588, had heard Ricci's science extolled and had sent to him his disciple, (Leuiuencien ?) " to learn from it the mistakes of Chinese mathematics." The latter found the Father at Peking and, about the month of June 1601, put Ricci in connexion with his master whom evil-doing eunuchs had caused to be imprisoned there. This distinguished man, during his three years of semi-captivity, reprinted " the treatise on the four elements with other mathematical questions, two small maps of the world and all that he had

[18] In our work "*Matthieu Ricci et la société chinoise de son temps*," we shall describe this dramatic episode in detail.

[19] R, I, pp. 391 *seq.*

been able to collect of the knowledge which had come from the West; he even prepared a translation of the great tables of the celestial constellations and would have had them engraved if he had not been prevented by certain considerations," doubtless the fear of appearing to encroach on the jealously guarded domain of the King's mathematicians.

Li Wo-ts'un, 李我存, later known by the name of Doctor Leo, was himself also a native of Chêkiang (in the neighbourhood of Hangchow). During his youth,[20] he had drawn up a Description of China with maps of the fifteen provinces, and he believed that the whole world was limited to that, but, when he saw Ricci's Map of the World in two Hemispheres:

> He[21]. . . was convinced that the correspondence between the celestial circles and the terrestrial degrees of latitude was based on an immutable law. As the result of most painstaking calculations which absorbed him uninterruptedly for a whole year, he was fain to acknowledge the narrow scope of his own previous map, which did not cover a tenth part of the globe as shown in the maps brought from the west.

From then he pressed Ricci to republish the World Map, in a considerably larger size; it was he, probably, who had it engraved on wood, because we are told that at that time, he occupied all the people of his palace in constructing many fine instruments, among others all kinds of sun-dials, described by Father Clavius in his Gnomonics, and a very fine astrolabe with its sights. The final edition of the Map of the World appeared during the year 1601; astronomy occupied a large part of it.[22]

At the right-hand corner, we first see a diagram representing the nine celestial spheres[23] and the times of their revolutions, with explanatory notes[24] on the left; then, proceeding down-

[20] R, I, p. 394.
[21] Giles, p. 368.
[22] Giles, pp. 372-377.
[23] Clavius, *Sphæra*, pp. 42-46.
[24] Clavius, *Sphæra*, pp. 55-57.

wards, we come to a long inscription signed by Ricci, in which he discusses the shape[25] and size[26] of the earth. He proves the fact of its sphericity by reference to his own experience at the Cape of Good Hope, where "I still had the heavens above my head and not beneath my feet." He enumerates the five zones and the five continents. . . .

Then follows an explanation of the parallels and meridians.[27] . . . On the left of the above is a "Short Discourse on the Four Elements"[28]. . . .

Below the Jesuit monogram at about "4 o'clock" in the circumference of the map, Ricci quotes a passage from the "Yüan History" (Chap. 52 *ad fin.*) on the varying lengths of days and nights, which shows considerable scientific knowledge for the fourteenth century. . . .

On the continent of Magellanica near South America is a table thus explained by Ricci:

Every degree of latitude, south and north of the equator, equals 250 *li* on the earth's surface. With regard to the longitude running east and west, only on the equator is a degree equal to this. In calculating the rest, the further you go north or south, the smaller the degree; so that these degrees of longitude are all less than 250 *li* in length.

Once more, according to Ricci's own opinion:[29]

They speak only truths which to us are banal and known to all, but for these scholars, they were the most profound truths, the most subtile and the newest which they had heard on these subjects; the greatest scholars of antiquity had fallen into manifest errors through having ignored the sciences and observations of our philosophers. That is why many people avowed, and still avow to-day that, thanks to these explanations, their eyes have been opened in very important matters in which up till then they had been as if blind.

What added value to these brief notices, for the most part extracts from the Treatise by Father Clavius on the Sphere of Johannes de Sacrobosco, is that Ricci had reduced many of the distances to Chinese measures.

---

[25] CLAVIUS, *Sphæra*, pp. 110-134.
[26] CLAVIUS, *Sphæra*, pp. 198-213.
[27] CLAVIUS, *Sphæra*, pp. 276-294.
[28] CLAVIUS, *Sphæra*, pp. 33-39.
[29] R, I, pp. 454-455.

One degree of latitude equals 250 *li*, and every minute 4⅙ *li*. Distance[30] from the centre of the Earth:

To the surface of the Earth: 14,318.9 *li*.
To 1st or Moon sphere: 482,522 *li*.
To 2nd sphere (Mercury): 718,750 *li*.
To 3rd sphere (Venus): 2,400,681 *li*.
To 4th sphere (Sun): 16,055,690 *li*.
To 5th sphere (Mars): 27,412,100 *li*.
To 6th sphere (Jupiter): 125,769,584 *li*.
To 7th sphere (Saturn): 205,770,564 *li*.
To 8th sphere (fixed stars): 322,769,845 *li*.
To 9th sphere (Primum mobile): 647,338,690 *li*.

These spheres enclose one another like the coats of an onion. All are hard, transparent, and colourless like crystal, and sun, moon, and stars inhere in their substance like the knots in a plank of wood.

Some eclipses,—those of the moon of June 15 and of December 9, 1601, and that of the sun July 4, 1602, gave Ricci an opportunity to measure the longitude of Peking. Later the eclipse of the sun of May 11, 1603, established afresh the unrivalled reputation of the missionary; it had been announced by the Emperor's mathematicians with some considerable errors: nearly three quarters of an hour (42′ 36″) for the beginning and about a quarter of an hour (14′ 12″) for the end.[31] Ricci, not being in possession of astronomical books, had to content himself with the ephemerides and tables which the Portuguese published for the use of the general public:[32]

Often, he wrote,[33] I predict eclipses much more exactly than the Chinese; that is why, when I say that I have no books and that I do not wish to apply myself to reforming their rules of calculation, there are many who believe me.

He also transposed the Gregorian Calendar, adjusting it to the lunar calendar, in such a way that the Christians

---

[30] Clavius, *Sphæra*, pp. 198-214.
[31] Souciet, III, p. 365.
[32] For example André de Avelar, *Repertorio dos tempos conforme a nova reformaçao do Santo Papa Gregorio XIII*; there are editions of 1585, 1590, 1594, 1602.
[33] R. II, p. 285.

PRIMA PARS

Schema huius præmissæ diuisionis Sphærarum.

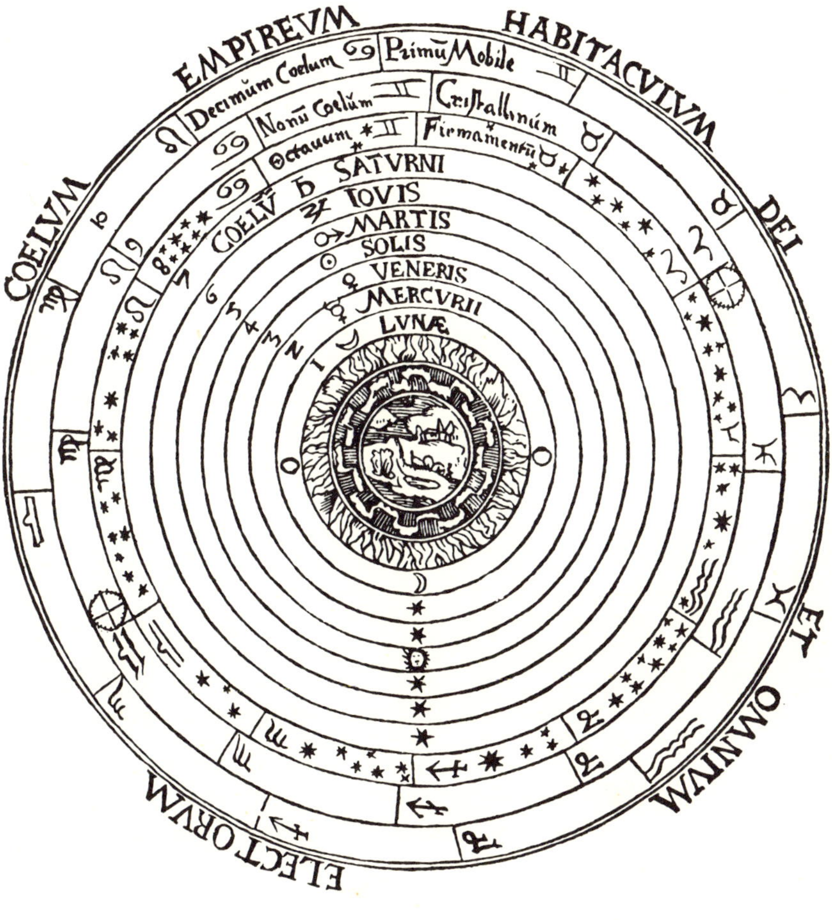

GEMMA FRISIUS, COSMOGRAPHICUS LIBER PETRI APIANI

Folio II, verso. (Library of the Old Jesuits).

could know all the movable and fixed feasts, as well as the lunations and the twenty-four annual periods :

All that is more exact[34] than the calendar which they compile each year at enormous expense ; the heathen are perplexed about it, because I explained it to them very clearly. Some even wished to print it ; I did not give my permission because in China one is exposed to very grave suspicion if one publishes new calendars.

From that time, he could outline with greater precision the true state of the astronomical sciences.

The Chinese, he wrote,[35] formerly possessed many notions on astronomy and the other mathematical sciences ; it was in arithmetic and geometry that they succeeded better. Their celestial constellations are not the same as ours, and they enumerate four hundred more stars than our astronomers do, because they add those that do not shine in a permanent manner [the comets]. But they take no trouble to explain the phenomena or the appearances ; they confine themselves to calculating as best they can the eclipses and the movements of the planets, and in that they make many mistakes.

That which especially occupies them, is judicial astronomy [or astrology] because they believe that all the events of this lower world depend upon the stars.

Only in the mathematical sciences, they have been aided to a certain extent by the mathematicians who came from Persia ; but the latter taught nothing by means of demonstration, they contented themselves with transmitting the tables according to which the Chinese calculate their year, the eclipses of the sun and moon and also the movements of the planets.

The founder of the [Ming] dynasty allowed only those whose duty it was [to fix the calendar], to study that science [astronomy], because he feared that rebellions would be hatched under its cloak. Also [the Emperor] maintains many astronomers in the interior of the palace, and these are all eunuchs ; other mathematicians reside outside the palace. He spends many thousands of crowns in salaries for these people, according to the rank which they obtained in their examinations. All these mathematicians are divided into two Boards or Colleges, both those who are in the palace and those who are outside

---

[34] R. II, p. 264.
[35] R. I, pp. 22-23.

it: the former calculate according to the ancient methods of China, the others follow the new method which came from Persia, and then those from inside and those from outside meet to compare their results and to aid each other. Each of these Boards has its platform or its tower in an elevated place, and they have made there some very fine mathematical instruments of bronze, extraordinarily large and very old, to observe the stars; each night, an astronomer remains there continuously to see if a comet or new object appears in the sky, and on the following day the King is notified of it in a public memorial the significance of the celestial phenomenon being explained at the same time.

Years pass; the Jesuit mathematician whom Ricci had requested be sent from Europe had not yet been granted:

I still insist, he wrote on May 12, 1605,[36] that they send us an astronomer... Here I have not enough books...The Board of Mathematics which follows the Chinese rules is more esteemed, but it deceives itself still more grossly; that of the Mussulmans is more contemned, but it predicts the eclipses with greater exactitude.. Besides the two Boards outside the palace and the two others of the eunuchs inside it, there are two at Nanking with many mandarins...I wish that this question be taken up again with our Father [General] as a very important matter for China. Never mind the nationality! Let one or two astronomers be sent directly to China, and even to Peking, for they would be of little use in another place. Let them take care to bring with them the necessary books, without relying on Goa, or other places: in fact, in the colleges in India, these books do not exist, and if there were any, they would not part with them to any other house.

Being unable to do more, Ricci confined himself to developing what he had formerly begun so favourably with Ch'ü T'ai-su at Chao Chou and with his pupils at Nanking. Under his direction, Li Wo-ts'un took advantage of a long sojourn at Peking to study with much diligence and application the works of Father Clavius:

Yesterday, wrote Ricci on May 12, 1605, he sent me from his new residence [in Shantung, Ciaucin (?), 35° 30′ latitude] two stone sun-dials, the one plane, the other fixed on a wall, and,

---

36 R. II, pp. 284-285.

which complicated the problem, the wall was not oriented to face the South, but was aslant; thus, to make a still more surprising object, he constructed this inclined, or rather deviating dial.[37] This mandarin had studied with me for some years past at Peking many mathematical subjects. . The name of Father Clavius is already known to many Chinese.[38]

Another remarkable convert, a native of Kiangsu, Paul Hsü Kuang-ch'i, 徐光啟,[39] dreamt of nothing but of adding to the prestige of the Fathers in order to aid them in propagating Christianity:

He decided with Father Ricci[40] to translate one of our scientific books to show the scholars of this kingdom with what diligence ours carry on their researches and on what solid foundations they establish their proofs; through that, they come to understand that in matters of sacred religion, it was not lightly that we had decided to take the side we did. After having spoken of several books, Paul Hsü and the Father concluded that, for the time being, it would be best to translate the books of the Elements of Euclid: in fact, however much mathematics were esteemed in China, everything in them was stated without proof and, on the other hand, just when we wish to teach some thing or part in a scientific manner, we can do nothing without this book [of geometry], especially in view of the great clarity of its demonstrations.

First, he came to an arrangement with a poor scholar who applied himself to this translation and obtained his livelihood from it; then Paul Hsü remembered that formerly Ricci had said to him: "only an intellect like yours is capable of bringing such an enterprise to a good termination," and Paul tied himself down for a year and more to come daily to begin the study with the Father for three or four hours.

When he began to understand the subtlety and solidity of this book, he took such a liking to it that he could not speak

---

[37] CLAVIUS, *Gnomonices* . . ., all of the third book.

[38] In Chinese, the name of Father Clavius is "Ting," 丁, the meaning of which is the same as that of the German word, "nail."

[39] We have sketched the rôle of Paul Hsü in the *Dossiers de la Commission synodale*, Peking, 1933, tome VI, pp. 941-952. (*Un précurseur de la Chine moderne, Paul Siu Koang-k'i*).

[40] R, I, pp. 500-502.

of any other subject with his fellow scholars, and he worked day and night to translate it in a clear, firm and elegant style. . . Thus he succeeded in reaching the end of the first six books which are the most necessary [Plane Geometry] and, whilst studying them, he mingled with them other questions in mathematics. He would have wished to continue to the end of the Geometry; but the Father being desirous of devoting his time to more properly religious matters and to rein him in a bit told him to wait until they had seen from experience how the Chinese scholars received these first books, before translating the others.

Toward the end of 1607,[41] the work was printed and it at once excited great curiosity in the intellectual circles of Peking.

It was admired rather than understood. It thus came very opportunely to lower the pride of the Chinese, for the best scholars had to admit that they had read a book printed in their language without being able to understand the meaning, even after applying themselves to it unremittingly; this being perhaps the first time that this had happened to them.

The Elements of Euclid thenceforth served as a foundation for the teaching of Ricci and Paul Hsü who both recruited some pupils, and, probably for the first time, the Father explained the complicated system of epicycles and eccentrics,[42] by which the astronomers of the beginning of the 17th century strove to give a rational account of the apparent phenomena of the heavens. An eclipse of the sun had been incorrectly announced by the official astronomers for February 26, 1607;[43] and this gross blunder had evidently done nothing more than excite the curiosity of those who interested themselves in the exact sciences. Thus, little by little, through the energetic impulsion of Ricci, a kind of centre of scientific culture, an embryo university or at least the equivalent of what we call a university extension course, was organized in the very capital of China.

[41] R, II, pp. 334, 342, 359, 372.—Pfister, No. 11, pp. 37-38.
[42] R, I, p. 311.—Clavius, *Sphæra*, 1585, pp. 431-458.
[43] Souciet, III, p. 365.

教者亦無所措其口也吾西庠如向所云幾何之屬幾
百家爲書無慮萬卷皆以此書爲基每立一義卽引爲
證據焉用他書證者必標其名用此書證者直云某卷
某題而已視爲幾何家之日用飲食也至今世又復崛
起一名士爲竇所從學幾何之本師曰丁先生開廓此
道益多著述竇昔游西海所過名邦每遘顓門名家輒
言後世不可知若今世以前則丁先生之於幾何無兩
也先生於此書覃精已久既爲之集解又復推求續補
凡二卷與元書都爲十五卷又每卷之中因其義類各

LEAF FROM THE PREFACE TO EUCLID WRITTEN BY RICCI IN 1607

(Note the name of Clavius 丁 in the third column from the left. See note 38 on page 67).

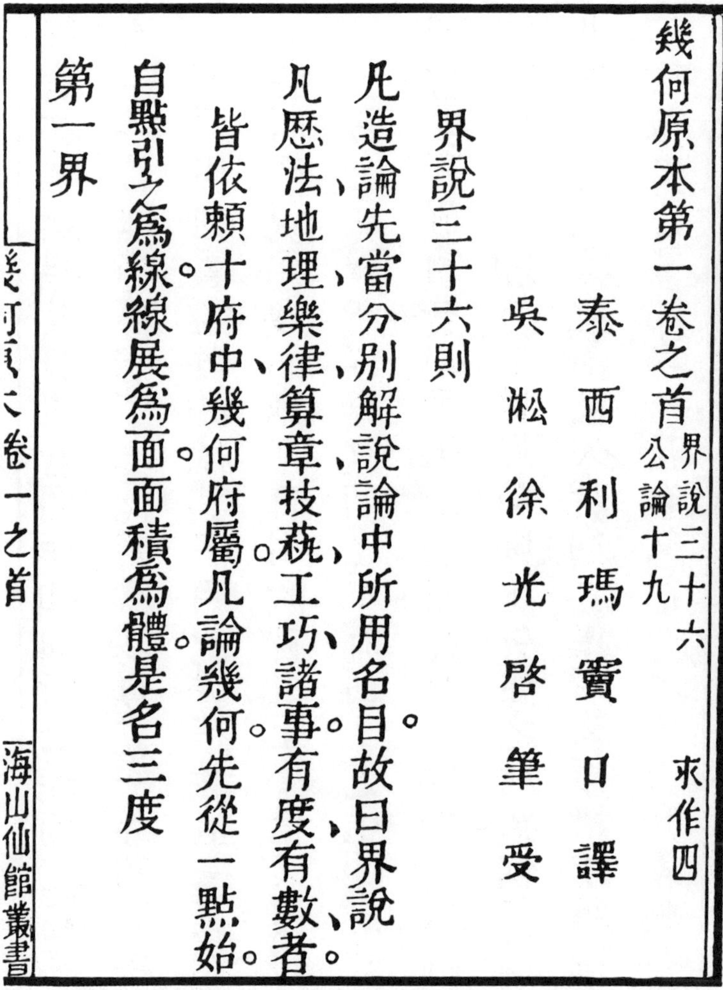
幾何原本第一卷之首　界說三十六　公論十九　求作四

泰西利瑪竇口譯

吳淞徐光啓筆受

界說三十六則

凡造論先當分別解說論中所用名目。故曰界說。

凡歷法、地理、樂律、算章、技藝、工巧、諸事。有度、有數者。

皆依賴十府中、幾何府屬。凡論幾何。先從一點始。

自點引之爲線。線展爲面。面積爲體。是名三度。

第一界

幾何原本卷一之首　海山仙館叢書

First Page of Ricci's Translation of Euclid

"The westerner Matteo Ricci 利瑪竇 orally translating, Hsü Kuang-ch'i recording with his brush."

A second centre also was born in the region of the Yangtzŭ River delta: Ch'ü T'ai-su continued to gather together his friends at Soochow; Paul Hsü, recalled to Shanghai for the funeral rites of his father, profited by his passage through Nanking and the enforced leisure of his mourning, to spread the doctrine of his master; Li Wo-ts'un especially began to publish quite a collection of scientific books at Hangchow whence he continued to correspond assiduously with Ricci. The alternatives of his official career brought him back to Peking several times; and he eagerly took advantage of them to extend his scientific knowledge.

In 1607-8, there appeared the abridgment of the treatise on the Astrolabe in two volumes,[44] as well as the treatise on the Sphere.[45] When he returned to the capital in 1608,[46] he prepared the "Complete Practical Arithmetic,"[47] adding to it "the extraction of square roots, cube root and cubic cube root in infinitum."[48] When he had mastered the first six books of Euclid translated by Paul Hsü, he completed his own treatise on the Sphere by that of isoperimetrical figures[49] and that of celestial constellations;[50] and used them to construct a celestial globe and a terrestrial globe. He also made a fair copy of his manuscript on Gnomonics.[51]

In these translations, Ricci and his collaborators did not servilely transcribe the works of Father Clavius, but

[44] There is here a slight contradiction between the letter of March 6, 1608, which speaks of "the past year" (R, II, p. 334: perhaps it is a question of the lunar year which finished on January 31, 1608!) and that of August 22, 1608, called "this year" (R. II, p. 363).

[45] R, II, p. 342: the letter of March 8, says "recently."

[46] Before August 22, 1608: R, II, p. 359.

[47] R, I, p. 396.—Pfister, 38, No. 12.

[48] Clavius, *Geometrica practica*, 1604, liber VI, problemata 14-16, propositiones 19-21.

[49] R, I, p. 396: according to the note of Father Tacchi-Venturi, one can believe that the question here is of the Spheres of Theodosus, but it appears beyond doubt that one must understand the treatise of Clavius on the Sphere of Sacrobosco.

[50] R, II, p. 547.

[51] R, I, p. 395.—We willingly think that the *De horologiis* spoken of in R, II, p. 363, is not different from this edition of the Gnomonics.

they took care to curtail many of the fastidious amplifications and prolix discourses with which the professor of the Collegio Romano compiled the Italian science of the time. Sometimes also, they added, or rather, they transposed a fragment from another work, as at the end of the third book of the Geometry of Euclid,[52] "the relation of one circumference to another of which the diameter is double, and the quadrature of the circle." [53] Finally, they did not cease reviewing and improving the first editions: as witness the copy of the Geometry corrected in Ricci's handwriting which perished during the storm of the summer of 1611.[54]

We are approaching the end of the career of the founder of the mission; his renown had extended beyond the frontiers of China. Uninterrupted relations united, through the intermediary of Macao, the missionaries in Japan with those on the continent; thus Ricci utilized the pictures which the Japanese Brothers painted under the direction of Brother Nicolas.[55] A scientific movement had taken shape in Japan, but it was less intense than that in China: the chief initiator seems to have been the physician Manase Dosan, baptised in December 1584, whose disciples rose to the number of 800.[56]

The World-map and the Calendar of Ricci were welcome[57] when in May 1605, an academy for princes was opened at Meaco: "They taught there geography, the use of the astrolabe and the movement of the planets." [58] The antichristian persecution has so thoroughly effaced in Japan the traces of religious propagation that it is very difficult to-day to be sure if the scientific treatises of Ricci

[52] R, II, p. 334.

[53] This demonstration of Archimedes is given in CLAVIUS, *Geometria practica*, 1604, Book IV, chap. VI.

[54] R, II, p. 546, note 11.

[55] *Artibus Asiæ*, 1930-1932, p. 67 (biographical data of P. SCHURHAMMER) R, I, p. 648, note I.—PFISTER, p. 124.—H. BERNARD, *l'Art chrétien en Chine du temps du P. Matthieu Ricci* (*Revue d'Histoire des Missions*, Paris, Avril 1935).

[56] MURDOCH AND YAMAGATA, *History of Japan*, vol. II, pp. 211-212.

[57] May 9, 1605: R, II, pp. 264-265,—Cf. R, I, p. 317.

[58] BARTOLI, *Il Giappone*, Turin 1825, vol. 3, chap. 26, pp. 130-131.

penetrated there;[59] it is not impossible, it is even probable,[60] but the edict of banishment of 1614 must soon have interrupted the current of influences which was only resumed at the end of the Ming dynasty.[61]

Until the end, Ricci insisted more than ever on the advantage of the great credit that the teaching of the sciences gave in China:

Although up to the present, he alleges,[62] I have taught them nothing else but a little mathematics and cosmography, they are so grateful to me that, often with my own ears, I have heard it said by distinguished persons that we have opened the eyes of the blind Chinese.

Let it then be decided, he insists,[63] to send us many books on the mathematical sciences and also others, and above all a good mathematician, especially an astronomer, who can continue what I have begun with my small energies, my few books and my little knowledge!

Among those "other" sciences that Ricci brought to China, was Architecture: some weeks before his death, he built a small chapel, reproducing the essential traits of European edifices.

In order that it may not resemble the temples for idols, he explained[64] . . . it was constructed according to all the rules of our churches: 70 hands long and 35 wide. The rules of architecture were observed when possible, it being granted that the workers had no notion whatever of them. All was made after the European manner with frontispiece, arches, cornices and lintels. . . In twenty days, the main part of the work was finished; but if it greatly pleased the Chinese who

---

[59] *Transactions of the Royal Asiatic Society of Japan*, 1913, vol. 41, part III, p. 383.

[60] Smith and Mikami, *A History of Japanese Mathematics*, 1914. pp. 214-215. —According to these two authors, there is no trace of Euclid's geometry in Japan before the 18th century, but they made little use of the Letters of the Jesuits.

[61] Kaempfer, *The History of Japan*, translated by Schechzer, Glasgow, 1906, vol. II, ch. IX, pp. 250, 255: the Japanese continued to organize a very strict bureau of censure to control the import of Chinese books dealing with the sciences or philosophy.

[62] R, II, pp. 384-385: letter of February 15, 1609.

[63] R, II, p. 343: letter of March 8, 1608.

[64] R, II, p. 613, note 2.—H. Bernard, *l'Art chrétien*. . . . (See note 55)

came in great numbers to see it, it only distantly recalled to us our European churches.

This was, one may say, the supreme gift of Father Ricci to China: the roof of the chapel was not yet covered in when his death suddenly occurred (May 11, 1610) at the age of 57 years.[65] Several times, in official circles in Peking, it had been considered to confide to him the revision of the calendar;[66] this project only began to be put into execution eight months later, but, by continuing to the end to initiate into European mathematics four or five fervent disciples, the founder of the mission deserves to be inscribed in the first rank of the reformers of Chinese astronomy.

---

[65] H. BERNARD, *Aux origines du cimetière de Chala. Le don princier de la Chine au P. Ricci (1610-1611)*, Tientsin, 1934, p. 52.

[66] BARTOLI, *La Cina*, Turin, 1825, Book 3, ch. 5, p.10.

CHAPTER V

THE SOLAR ECLIPSE OF DECEMBER 15, 1610

On December 15, 1610, following the immemorial prescriptions of Chinese law,[1] all the officials of Peking, in full dress, had gathered together with the bonzes, the Taoist monks and an immense crowd for the important operation of " saving the sun "[2] during its eclipse ; it was necessary, at the cost of a deafening hubbub of various instruments and with repeated profound bowings or prostrations, to frighten the dragon which was supposed to be devouring the star ![3] At 2 o'clock 12′ 14″, the moon should have begun to interpose itself between the earth and the sun, but the indentation of the luminous globe was only observed 30′ 22″ after the time announced by the Emperor's mathematicians ; the middle of the phenomenon, which had been predicted for 3 o'clock 42′ 36″, took place at 4 o'clock ; and as to the end of the eclipse, fixed for 5 o'clock, it could not even be ascertained for the sun had then already set ![4]

The scandal, this time, was immense,[5] and the kindest interpretations were incapable of excusing the blunder of the official astronomers. One mandarin, of fairly high rank, who had carefully observed the eclipse, denounced this crime of lese-majesty in a memorial addressed to the throne, and he published, on this subject, a mathematical work, " in which his good intentions were more evident

---

1 HOANG, *Mélanges sur l'administration*, Shanghai, 1902, pp. 91-93 : Detailed description of this official " rescuing " ceremonial.

2 Fathers Trigault and Bartoli say " of the moon : " this is certainly erroneous.

3 R, I, p. 23.

4 SOUCIET, III, p. 366.

5 TRIGAULT, pp. 119 *seq.* ; BARTOLI, *La Cina*,, Turin edition 1825, l.3, ch.5. pp. 9 *seq.* summarizes it fairly accurately.

than his knowledge." The mathematicians were severely reprimanded and punished, but they cleverly replied that they were punished for the faults of others since they were obliged to apply their astronomical tables; it was not they who were guilty, but their predecessors, and, if their books were not corrected, the same errors would continue to be reproduced. To this reply, nothing could be gainsaid, and the supreme Board of Rites, on whom depended this question of capital interest for the very existence of the Empire, had to examine the problem and find its solution.

Ricci had been dead for about seven months, but his science survived him in his companions and disciples. Two Fathers were then residing at Peking. Father de Pantoja, a Spaniard, who devoted himself to apostolic ministrations, had received a summary initiation in applied mathematics during nine years of sojourn with Father Ricci: he could play the harpsichord a little, repair clocks, observe latitudes, construct sun-dials, ...[6] The scientific training of the second Jesuit, Father Sabbatino de Ursis,[7] was more advanced. The latter had only arrived in Peking in 1607, he had received his education in philosophy at the Collegio Romano before joining the Society and he had studied mathematics for a year at Goa (about 1601-1602).[8] Although quite new to the mission, his kindly nature, his humility and his spirit of mortification had marked him out in Ricci's mind as his successor at the head of the mission in Peking: more than Pantoja, he was also qualified to accomplish the scientific work of the founder. For some months in which discussions developed between the imperial astronomers and their accusers, the Fathers were often consulted by both parties; the mathematicians even requested them to correct the Chinese astronomy and to translate the European books.

The occasion was too favourable: Paul Hsü resumed some negotiations which had miscarried several times. Somewhat afraid that the missionaries' science would not

---

[6] PFISTER, No. 19, p. 69.
[7] *Ib.*, No. 30, p. 103.
[8] R, II, LVIII-LIX and note.

answer the general expectations, he intervened with the Board of Rites requesting that a petition be addressed to the Emperor, in order that this most urgent reform might be entrusted to them.[9] In order to remain within all formalities, he requested moreover that the mandarin who had accused the official functionaries should be joined with them; an old scholar was also designated to help them. The parleys ended successfully, but the initial project was slightly modified: the Fathers, being foreigners, were only nominated to assist Paul Hsü Kuang-ch'i and Leo Li Wo-ts'un with their advice; a particule, inserted in the text of the memorial, implicitly authorized them to recruit other fellow-workers from their country. The imperial authorization was granted, and this sensational news was immediately made known throughout China by the official Court Gazette.

At first, everything seemed happily arranged for the success of the enterprise: the scholar who had accused the astronomers had ruined his credit by his scientific errors and besides, died soon afterwards; the old mandarin who had been associated with him refused all collaboration through his contemptuous arrogance. There remained then only the Fathers, especially Father de Ursis, with Paul Hsü to apply himself effectually to the reform of the astronomy. Unfortunately, the mathematicians unsettled all the plans through their indiscreet exactions. According to custom, in fact, after the Emperor sanctioned a memorial, a second petition should determine the manner in which his orders should be executed; the members of the Board of Mathematics, emboldened by their first success, increased their demands: it was necessary, they said, to repair their dwellings, construct new instruments, grant them profitable offices, etc. The Emperor left their request unanswered; they returned to the charge a second and a third time, so thoroughly that they tired the patience of the monarch and the problem of the reform of the calendar was adjourned. Probably also, resistance had had time to

[9] TRIGAULT, pp. 121-122.

organize itself against the intrusion of foreigners in a question so intimately connected with the very essence of Chinese civilization; the anti-foreign scholars preferred the methods of the Mussulman Board of Mathematics, already acclimatized in China, to the innovations of the "Western doctors." [10]

However, Father de Ursis had not waited for the conclusion of these negotiations to set to work; with the co-operation of Paul Hsü, he had begun without delay to examine the Chinese astronomical books [11] and, in April 1611,[12] the president of the Board of Mathematics came to learn from him the theory of the apparent movements of the sun and the moon. Father de Ursis then drew up an important programme of reform, of which we know the main points because he wrote about them to his superior, Father Pasio, in 1612.[13] The original text is in Portuguese and only came to us in a Latin translation; [14] since it is, it would seem, the oldest document in which Chinese astronomy is critically examined according to the data of the European science of that time, it deserves a methodical and detailed analysis.

In a general way, one may say that Father de Ursis, writing at an epoch when the question of origins was hardly ever thought of, was like all his contemporaries imbued with the Greek method, essentially ecliptic, that is to say that unconsciously, he was persuaded (like most of the historians of Chinese astronomy) [15] that the ecliptic is the necessary basis of astronomy and of the calendar. This reservation once admitted, it remains true that his modest exposé indicates an enquiring and honest mind.

---

[10] This is what took place in 1623: WIEGER, *China*, p. 250.

[11] WIEGER, pp. 249, 487: probably, they consulted the work *Shêng-shou wan-nien li* (published in 1595) by Chu Tsai-yü, a complete treatise on the calculation of time with a perpetual calendar, a masterpiece which ......magisterially sums up all the previous works on the subject.—SOUCIET, II, pp. XIII-XIV.

[12] BARTOLI, *Cina*, Turin, 1825, L.3 ch.5, p. 11.

[13] PFISTER, No. 8, pp. 24-25.

[14] TRIGAULT, pp. 124-138.

[15] SAUSSURE, *Origines*, p. 8.

After a short historical introduction,[16] he studies successively the theory of the calendar[17] and of the ephemerides,[18] concluding briefly on the way to correct them:[19] for several pages, he throws some light on a subject which the professional astronomers themselves approach with repugnance, on account of its complicated nature.

The historical introduction first cites the text, traditionally attributed to the Emperor Yao, in which are set forth astronomical notions going back "to 2358 B.C.," says Father de Ursis following the conventional chronology[20] without appearing to suspect the problems of concordance with the Vulgate Translation of the Bible which could in his time lay down this single date. As has been written:[21]

The discovery of China by Europeans has been slow and progressive. It has not been a question, in fact, for China, as for America, of discovering a new continent. . . and if China had been a country inhabited by peoples of a primitive civilization, without tradition and without history, its discovery would assuredly have been less important than that of Australia, for example. But China was a civilized country, having a civilization quite different from the European civilization, which has evolved by virtue of its own laws: [almost] unknown to Europe, it had not known [much about] it, and, as Europe herself believed, she was persuaded that she was the centre of the world. The civilization, which was born in the countries bordering on the Mediterranean and which had prospered so magnificently, was then not the only one which existed. There was another in Farthest Asia, which yielded in nothing to the Greco-Latin civilization, not only through its perfection, but what was more important, through its antiquity. . One cannot but stress the extent to which this idea of China's antiquity. . . attracted or repelled, in any case set going, the [European] spirit in the 17th century. . . Its theological consequences. . . are not negligible, far from it. If this chronology was fully

---

[16] TRIGAULT, pp. 124-127.
[17] *Ib.*, pp. 128-132.
[18] *Ib.*, pp. 132-136.
[19] *Ib.*, pp. 136-138.
[20] WIEGER, p. 10, who adopts the traditional chronology says: 2145.
[21] PINOT, *La Chine et la formation de l'esprit philosophique en France*, Paris 1932, pp. 108-119.

proved, it ran the risk of rendering suspect, or even of destroying the chronology [admitted according to the Vulgate] of the Bible, for it went beyond it.

Neither Father de Ursis, nor Father Ricci who mentions the still more distant date of 2636 B.C. in relation to the invention of silk,[22] was troubled by a problem which impassioned the intellectuals of Europe. Very simply also, Ursis speaks of the eclipse which occurred under the Emperor Yao without foreseeing the animated discussions which it had provoked and would, probably, still provoke for a long time to come.[23]

After the Burning of the Books, Ursis attributed to the Emperor Wu of the Han dynasty (104 B.C.), the restoration of astronomy.[24]

From that time, he adds in a more precise manner than Ricci ever wrote, the Chinese divided their mathematical sciences into two classes.

1.—The *t'ien wên*, 天文, which means 'the ornament of Heaven' and is concerned only with divination.

2.—The *li fa*, 歷法, that is the 'rule of order,' including all that we understand by theoretical or practical astronomy.

The first is officially reserved to the mathematicians only of the Royal College; the second is forbidden to no one, and that is why many scholars joined the school of our [mathematicians] to learn it, especially Paul and Leo. . .

However, the royal mathematicians are commissioned to take up both: their whole occupation reduces itself to predicting eclipses, drawing up the calendar, and observing the heavens by day and night; if they discover there a new phenomenon, they have to notify the King of it and explain its auspicious or inauspicious signification. Excepting them, it is prohibited to all, under the gravest penalties, to publish a calendar or an explanation of celestial prodigies; no one can, on his own account, even observe the stars.

---

[22] R, I, p. 4; WIEGER, *Textes historiques*, p. 25.

[23] SAUSSURE, *Origines*, pp. 1 *seq.*: he speaks appropriately of the "wind of madness" which was passing over certain astronomers.

[24] Not without Western influence, it seems: WIEGER, pp. 84-85; H. MASPERO, *Chine antique*, pp. 614-620.

In course of time, through not properly understanding the essential distinction between *t'ien wên* and *li fu*, many of the Jesuits in China and out of China asked themselves if they were not collaborating in a superstitious act by helping to draw up the Chinese ephemerides; they decided this matter of conscience in the affirmative and made it a criminal act on the part of Fathers Schall and Verbiest to accept an office in the Board of Mathematics.[25]

Father de Ursis concludes his brief historical introduction by showing that the method officially adopted for fixing the Chinese calendar had in the course of centuries undergone fifty-five reforms, the last in date being that of Kuo Shou-ching. Conjointly with it, they employed another method, known as "the Mussulman," *Hui hui li*, and translated 230 years before, under the founder of the Ming dynasty.[26]

In the one as in the other, they only use numerical tables without any theory, and that, concludes Ursis, is the reason of Father Ricci's success.

These historical preambles once settled, Father de Ursis analyzes the Chinese calendar. This differs from the Gregorian calendar especially in not being perpetual: it was compiled each year. This fundamental difference entailed great simplifications: no epact (the age of the Moon on January 1), nor golden number (corresponding to the 19-years cycle of Meton), nor solar cycle (the series of 28 years after which the days of the week correspond in the same order with the days of the month), nor dominical letter (that of the seven first letters of the alphabet which should be appropriated to the first day of the perpetual calendar).

Consequently, to correct the Chinese calendar, it is not necessary to make use of all what, perhaps, has most complicated the Gregorian reform.

---

[25] VÄTH, *Johann Adam Schall von Bell*, Cologne 1933, pp. 267-294.

[26] WIEGER, p. 241: *Shou-shih li* inserted in chapters 52-53 of the History of the Yüan. Ricci knew of these chapters, for he printed an extract from them in his World Map, GILES, p. 374. *Cf.* again WIEGER, p. 250.

But could this annual calendar be corrected, and even made perpetual? Father de Ursis answered in the negative, and, to understand his reasons, we will criticize them by the help of the very luminous writings of M. de Saussure. In China, remarks de Ursis, the year is lunar.

There is not rightly speaking a lunar year, says M. de Saussure,[27] for one year of the moon would be its revolution, that is to say a lunation. In fact, all the methods of the year in use relate to the solar year which produces the course of the seasons; the adjectives 'lunar,' 'solar,' are relative to the manner of computing the period, and not to the existence of the year. . The expression 'lunar year' expresses only a duration. And owing to the sway of words over ideas, there has been retained from the lunar year only this duration, the essential question of the starting point being neglected. The primordial element of the calendar is the point of departure, that is to say the method serving to mark the beginning of the year.

Ursis felt how important it was to determine this point of departure, but, conforming to the established usage, he begins by explaining that the Chinese year is composed of twelve lunations (with an intercalary lunation added every five years approximately).[28]

These lunar months are moreover unequal, some of 30 days (great months), the others of 29 (small months), because the true lunar month is included between these two numbers (29 days 12 hours 44 minutes).

If from the year and the month, one passes to the day, one sees that it is "natural," that is to say speaking like our contemporaries, it is the true solar day and is counted from midnight to midnight.

The subdivision of the day is the hour, but the Chinese hour is equivalent to two European hours because there are only 12 Chinese hours in a day. Another subdivision of the day is the quadrant: neither is the Chinese quadrant equivalent to the European quarter of an hour, because in a day there are 100 Chinese quadrants in place of the European 96 quarters of an hour. Besides, the centesimal division is applied to the Chinese quadrant, to the Chinese

---

[27] SAUSSURE, *Origines*, p. 116.
[28] HAVRET, p. 16.

minute, and so on : a Chinese quadrant equals 100 Chinese minutes, whereas a European quarter of an hour is equivalent to 15 European minutes. Therefore it would be very complicated to reduce the time data of the Chinese to those of Europe.

The manner in which the Chinese proceed to fix the beginning of the year is characteristic :

The polar star, wrote Saussure three centuries later, is the concrete basis of Chinese astronomy.[29] The horizon, and then the oblique course followed by the moving luminaries, served as an obvious mark for the ecliptic method. But the equator is a purely ideal notion and it seems incredible at first sight that an equatorial astronomy could be constructed directly without passing through the zodiacal form . . . (And yet) just as zodiacal astronomy has the horizon for its primordial element, the natural mark which led it to the rational notion of the ecliptic ; so diurnal astronomy has the meridian for its primordial element, conceived as the vertical of the polar star, the natural mark which led it to the rational notion of the equator (the circumference of the vault of heaven).

Father de Ursis expresses himself more simply :

Around the pole as a centre, the Chinese imagine a great circle (that of the equator) like that which ours call the 'apparent circle' (of the horizon). They divide this circle into twelve parts or hours. Because that circle is, in their view, perfectly stationary, it is necessary for the hours always to remain in the same place, that is to-day midday above, midnight below, six hours of morning in the east, six hours of evening in the west.[30]

Why did the Chinese year begin successively at midnight, two hours after midnight, four hours after midnight ! Father de Ursis avows that he cannot tell exactly.[31] Actually, he adds, at the beginning of each month, the Chinese astronomers assign two characters, one for the year, the other for the hour, to say " this month corresponds to this hour;" at the beginning of the first month they

---

[29] SAUSSURE, *Origines*, p. 15.
[30] HAVRET, p. 32 : where one may find some interesting supplements on this question of the Chinese hour.
[31] *Ib.*, p. 21.

always join the character 寅 *yin*, which is that of four hours after midnight.

How did they fix the beginning of the year?

They determined in the zodiac the sign called *li ch'un*, 立春, that is "spring begins," which is distant one month and a-half in relation to the spring equinox and which falls on February 5. The new moon which is nearest to it is, conventionally, the beginning of the year. That is why it is important to observe exactly, on the evening of February 5, in which region of the sky are the stars of the tail of the Great Bear.

De Ursis then adds a remark which conveys much on the causes of the inextricable confusion in which Chinese astronomy then was:

It is certain, nevertheless, that the Chinese will not change their method of beginning the year, even if the stars (of the Great Bear) did not to-day occupy the place which they imagine.

Of the sexagesimal cycle (and not secular, as in Europe) in which the years succeed each other,[32] the Father gives a fairly precise idea, and he shows that this cycle applies also to days[33] and to hours:[34] "in this correspondence, they are persuaded that there are profound mysteries, and they try to persuade others accordingly."

Up till now de Ursis has not spoken of the week, but, as in the sexagesimal cycle of the Chinese he recognizes a certain resemblance with the golden number of Europeans, he notices the analogy of the solar cycle and of the dominical letter with the Chinese *su*, 宿, of which Biot has since explained the origin, using the notes of Father Gaubil.[35]

As the measuring of the intervals of time are the more difficult and the more subject to error the longer they are extended, the Chinese had imagined, to render it surer and more convenient, a method that we ourselves use. They had chosen certain stars the number of which was definitely twenty-eight, divided in a quite unequal and apparently whimsical manner, on

---

[32] *Ib.*, p. 1.
[33] *Ib.*, p. 21.
[34] *Ib.*, p. 32.
[35] Cited by Rey, *La science orientale avant les Grecs*, p. 364.

the whole contour of the heavens. They measured as exactly as possible the intervals of time which elapsed between the successive passages of these fundamental stars over the meridian. Then, when they wished to determine the relative position of any other star in relation to the diurnal movement of the sky, they had only to observe the interval of time which elapsed between its passage over the meridian and that of the nearest fundamental star. Thus they always expressed the places of the stars by this interval converted into an arc.

The missionaries Terrenz, Rho, Schall (1629) later accepted this system and knew how to obtain from it the calculations for their astronomy. For the time being, de Ursis only wished to extract from it what concerned the problem of the week.

Each of these *su* [he says : 'constellations,' as does Ricci elsewhere,[36] and this translation implies quite an exegesis of the astronomical system of the Chinese] has its own character, and each of these characters combines one by one with those of the planets. Now as the planets are seven in number [for the Chinese as for the Europeans of the beginning of the 17th century, the sun is one of the seven planets], each of them appropriates four characters of the *su* [the number twenty-eight is exactly divisible by seven], and that exactly corresponds to our Solar Cycle : . . and because in China there is no intercalary day [for the bissextile year], it is impossible that the series of the characters of the *su* should not repeat themselves in an unchangeable manner for each day [of four in four weeks]. It sufficed therefore for our Christians to look in the Chinese Calendar for the four characters appropriated to the sun to know when the Sunday took place. . ; for the other feasts, we print a calendar.

From this account, it is clear that the Chinese calendar, so different as it at first sight appears from the European calendar, is perfectly logical and coherent ; it is even not very difficult to make them agree if one observes that four of our weeks are strictly equal in duration to the series of days to which are assigned the characters of the twenty-eight *su*, and one can adopt the conclusion of Saussure :[37]

---

[36] R, I, p. 316, line 36 . . . . . . R, I, p. 22 ; II, p. 216.
[37] SAUSSURE, *Archives*, p. 348.

It is an admirable system. It would be wrong to suppose that it implied an exact knowledge of the celestial movements and great precision in observations. . . but the notion of the ecliptic not having attracted attention and the philosophical conceptions compelling them to consider the movements of the mobile axes as influenced by moral causes, one would hardly attempt to perfect the observations.

In other words, if one only examines the Chinese astronomical system from the point of view of the calendar, no correction is called for; but if one attempts to make the calendar agree with the celestial phenomena, otherwise stated if one wishes to arrange the ephemerides, one encounters something greater than a difficulty, and it is that which Father de Ursis examines in the second part of his small memoir.

He begins by eliminating from his researches the enormous part given to astrological divination or to conjectural astronomy:

The Mathematicians combine their five elements with the characters for the hours and the *su*, then they compare the planets with them. . . and it is that wherein consists their astrology, almost all their philosophy of nature and all their medicine. In these matters, the Chinese have discovered nothing truly remarkable, and all their eminence comes to them from their moral philosophy and their faculty of correct expression.

De Saussure later on gives some examples of this astrology:[38]

The appearance of some constellation, at some epoch of the year, not only evoked an idea of concomitance, it was not considered simply from the point of view of its calendric utility: it awoke before all a (kind of) religious emotion; the heavens and the different parts of the heavens were considered as the immediate cause of the terrestrial transformations which mark the course of the seasons. If spring succeeded winter, it was not,—as we think to-day,—because the increasing height of the sun increases the intensity and duration of its calorific power, but because the dragon was substituting its influence (*yang*, 陽) for that of the tortoise (*yin*, 陰). And the progress

[38] SAUSSURE, *Origines*, p. 378.

of that substitution could be read, each evening, at dusk, on the eastern horizon.

The ephemerides include also, according to Father de Ursis, the new moons, the full moons and the quarter moons, the equinoxes, solstices and signs of the Zodiac, the lengths of days, the time of the rising and setting of the sun, the positions of the planets. It is here that the weakness of their astronomy revels itself.

Doubtless, further states Father de Ursis :

> The Chinese know the equator and the zodiac, but they are ignorant of their properties and use. They content themselves with affirming that the sun always moves in the middle of the zodiac. As to the moon, they assign to it nine routes, each of which has been given its own special name ; they differ according as the moon appears to enter or leave the ecliptic, to the North or the South, to the East or the West ; on these routes, they determine 363 points where an eclipse is possible, and it is on that that they make their calculations. They have never thought of dividing the equator, but they partition the zodiac into 365° 24,′ exactly according to the number of solar days, then they distribute these degrees [of the zodiac] in 24 equal parts which they call signs and each of the signs uniformly includes 15 days 21 quarters 84 minutes.

This material text contains a mistake which is poignantly exposed by de Saussure :[39]

When the Jesuits arrived in China, they found there two kinds of division in use :

1.—the 28 *su* serving to fix the horary interval of the sidereal positions in relation to the 28 determining stars, of which they were charged to indicate the arrangements for K'ang Hsi's encyclopædia (at the beginning of the 18th century).

2.—an equatorial division of the sphere into 12 *chung-ch'i*, 中氣, or temporarily equal parts serving principally to fix the rule of lunar-solar intercalation.

Fully impregnated with the Greek principles, they could not conceive the equatorial character of Chinese astronomy . . . Thus they attributed, through an erroneous assimilation,

[39] *Ib.*, p. 68.

the words zodiac, signs of the zodiac to the Chinese 12 *chung-ch'i*. (He would have been more correct had he said the 12 *ch'i*, 氣).

It remained to find a term to designate the other system of division, the *su*. Now, by analogy, it was obviously indicated to call them Constellations, since they called the *Chung-ch'i*, Signs. . . . That was quite logical. At no time must one lose sight of the fact that this assimilation admits of a difference: the Greek system relates to the ecliptic and the Chinese system to the equator.

Thus one sees what was the mistake made by de Ursis,—and, we must add, by Father Ricci, from the first period of his sojourn in China[40]—namely, that of transposing to the ecliptic the divisions of time which belong to the equator. One can excuse them when one thinks that, even in our day, after the luminous demonstrations of Father Gaubil, of Biot and of de Saussure, many historians of astronomy commit the same error;[41] the astronomers of the Mussulman Board of Mathematics were themselves also unaware that the Signs employed by them, being based on the ecliptic, differed from the Signs based on the equator which were used by their colleagues of the Chinese Board. Astronomical science in China had fallen into such a state of confusion that no one was capable of discerning its misconceptions. Besides, no harm resulted from this for the calendric reform of the Jesuits, since they concluded it was necessary to reject the Chinese tables and translate the European works such as they were.

To explain the inequalities of movement of the sun in the zodiac, the European scholars, partizans of Ptolemy's system, imagined a very complicated combination of eccentric and concentric circles. The Chinese were ignorant of them as well as of the apogee, and that is why, says Father de Ursis, "their signs of the zodiac cannot correspond to the European signs, nor the days when the

---

[40] November 24, 1585: R, II, p. 72, "I have found they have 12 (*ch'i*) or, to speak more correctly, 24 (*ch'i* and *ch'ieh ch'i*) times of the year which correspond exactly to ours." *Cf.* R, I, pp. 264-265; II, p. 205; II, pp. 241-242: The signs of his sun-dials.

[41] HAVRET, p. 113 (The 24 stations of the ecliptic) does not seem exempt from this confusion.

Sun enters those signs, with the exception of the equinoxes and the solstices!" This reason is not valid here, but that which is added later is of great importance: "The error we are called upon to correct, is the difference between our positions and theirs for these points of the equinoxes and the solstices."

There follows a series of remarks on details:

> Their division of degrees and minutes is centesimal [and not sexagesimal, like ours] . . . Though they usually count nine heavens, they do not mention them in their ephemerides. They entirely ignore the eccentrics, the concentrics, the epicycles, they have never heard of the proper movement of the planets fom West to East.
>
> Finally they do not know of the other phenomena observed by Europeans, in particular the parallax or difference of aspect [depending upon whether an observer is placed at the centre of the earth or on its surface] so necessary in astronomy, still much less refraction.

Father de Ursis noted exactly how the Chinese had established simple and convenient correlation to evaluate the progress of the sun in space and time, thanks to a conception analogous to our actual fiction of mean time:

> They fix generally and imagine a certain movement [of a fictitious sun] which is produced in the space of time elapsing between a winter solstice and the following solstice, and this space is precisely the same as what Europeans call the space of a year. Since a day is equal to 100 quadrants, a quadrant to 100 minutes, the day is therefore equal to 10,000 minutes, and, they affirm, this movement [mean] takes place in 3,652,575 minutes. They then determine for each planet [the duration of] its movements, and likewise for all the others: the [true] movement of the sun is comprised in 3,652,425 minutes.

In all these evaluations, they make no mention of the different heavens, nor do they explain the accelerations or retardations of their movements, and they ignore entirely the other properties of the planets: in a word, they do not know the theory; consequently:

> They are incapable of scientifically correcting, if I can so speak, their errors concerning the celestial movements. . . . and it is that in which the reform of the Chinese ephemerides consists.

Nevertheless, they remedy it practically by the following process: they set up, perpendicularly, a style eight cubits in height, graduated in degrees and minutes. They take the winter solstice for the point of departure of their observations, whereas Europeans have adopted the spring equinox; that is why three days before the winter solstice, they daily observe the shadow of that gnomon, and note its degrees and minutes; three days after the solstice, they continue these observations, and by comparing them they easily discover what is the day, hour and exact quadrant when the solstice began. This same operation is performed in different parts of the kingdom: they spare neither time nor trouble over it, they assign to it the highest magistrates of the capital and it is the principal point of their correction. For they devote the remainder of the time to reducing the movements of the planets according to this correction, and to observing the stars from an elevated place. . . . Among the fifty-five reforms of the Chinese calendar, the most celebrated is that of Kuo Shou-ching who not only determined the winter solstice, but also the different latitudes of places, hardly known [to the Chinese of to-day].

The substantial analysis of the Chinese ephemerides stops here, and it only remained for Father de Ursis to formulate his conclusions:

Since the Chinese make no use of the epact, of the golden number, of the solar cycle and the other elements found in our Calendar, the whole difficulty of the correction of their calendar consists in these two points:

1.—the precession of the equinoxes.

2.—exact knowledge of the movement [of the planets] in order to predict eclipses.

The precession of the equinoxes is known to them, not because the prime mover (the sun) is animated by two movements, the one which impels, the other which is proper to it and acts in a contrary direction, but because they have observed this variation of the day only in the winter equinox. Thus, they say, under the reign of the Emperor Wu Ti [of the Han dynasty], the sun had varied 24° since the epoch of their first king, in the interval of 2,000 years. When they speak of this difference, they do not attribute it to the stars, but to the solstice: at a certain epoch the solstice is in a certain constellation, and at another epoch it is in a different degree. They do not seek further to find out the reason. Several times, they have tried to evaluate this displacement. . . . the last time, the great mathematician Kuo Shou-ching gave it a value of a degree in sixty-six years,[42] and it is the figure which they have admitted up to the present. . .

As regards the movement of the planets, they know it, but, besides the fact that they do not avail themselves of the precession of the equinoxes, their tables are defective, so much so that they are mistaken by three or four quarters of an hour almost for each eclipse. This error cannot be attributed only to the diversity of aspect [caused by the parallax, reflexion. . .], for in the eclipses which take place at midday, when the difference of longitude is the minimum or even null, they err in the same way. One reads in their books,—although they do not speak in them of aberrations of aspect,— this rule : "If the eclipse of the sun takes place at midday, the [calculated] number must be subtracted; if it is after midday, it must be added," and they give no explanation. The mathematicians do not know the cause of this error : they confine themselves to saying that they have the tables, they do not know the foundations of them.

On this so astonishing absence of curiosity of the Chinese astronomers, it is useful to read a characteristic page of Saussure :[43]

The discovery of the precession well shows the difference which separates the Greek genius from the Chinese spirit. A single previous observation, that which Timocharis had made 122 years before him, sufficed for Hipparchus to understand that it was a question of a law of nature, to envisage the two hypotheses of the mobility of the ecliptic or of the equator, reject the first and approximately determine the conditions of the second. In China, on the contrary, . . the astronomers passively stated for more than twenty centuries the discordance between the real situation and that of the [Chinese] system, without thinking of modifying the system nor of analyzing the progressive displacement. They were content, according to the physico-moral determinism of their philosophy, to admit that the heavens were out of order, attributed this change to the decadence of morality and concluded therefrom that it was necessary to reverence the golden age of remote antiquity in

---

[42] SOUCIET, II, p. 108, says in 67 years ; III, pp. 147-148.
[43] SAUSSURE, *Archives*, p. 207, note 1.

which, thanks to the omniscience and sanctity of the Emperors, the Heaven and the Earth were in perfect concordance. It was only at the beginning of the modern era, under the Han, when they established century by century the continued retrogradation of the solstitial place, that the phenomenon was officially signalized to the Emperor, but a hundred years were still necessary before they saw in it a permanent law, and—an incredible thing—they continued to discuss for another six centuries before arriving at a satisfactory evaluation of the precession.

A passivity of the same kind stood in the way of the calendar being corrected at the beginning of the 17th century. However, says Father de Ursis :

Doctor Paul has gone over with me [the European theory of] the precession ; he shows an extraordinary liking for it and judges it very suitable for correcting the errors of previous observers.

As to the tables of the planets, the royal mathematicians and we ourselves, do not think of correcting them : that would be a very difficult undertaking on account of the [centesimal] division of the minutes in China which it would be very laborious to reduce to our [sexagesimal] method of calculation ; besides the mathematicians themselves greatly prefer our division of the minutes. Our intention then is to leave on one side the Chinese tables and substitute ours for them, and to establish the Chinese ephemerides starting from their data.[44]

To sum up, we will say with de Saussure : [45]

On account of the importance attached to the notion of the pole, the centre of the heaven, Chinese astronomy has from the beginning had this equatorial character which it preserved until the intervention of the Jesuits. No trace is to be found in it before the Han of the notion of the oblique circle. And even in the new era the divisions of the ecliptic are subordinated to those of the equator. In consequence of this habit of thought, when the Chinese discovered the law of precession, they interpreted it as equatorial, that is to say as if the centre of movement was the pole. They never suspected that the pole, symbol

---

[44] Father DE URSIS cites the *Alfonsine Tables* (TRIGAULT, 1611, p. 128) : Probably he had to use the *Tabulæ prutenicæ*, of ERASMUS REINHOLD (LIBRARY).

[45] SAUSSURE, *Origines*, p. 550.

of rectitude and immobility, could vary in the course of ages. They never doubted that the equator of remote antiquity was not the actual equator and did not in any way correspond to the actual pole.

Father de Ursis,—and his successors, Terrenz, Rho, and Schall,—did not think for a moment of making this tricky and, let us add, dangerous correction, for it profoundly shocked Chinese sentiment. He preferred to substitute the Greek and ecliptic method for the older equatorial principle, at the same time preserving the fundamental institutions of the Chinese astronomy and calendar, such as the divisions of the firmament and the tropic year.

CONCLUSION

WITH the arrest of the reform of the calendar by Paul Hsü Kuang-ch'i and Father de Ursis, we may say that the period of Ricci's contribution of European science to China comes to a close. During the following years, whilst the influence of the founder of the mission continued to be exerted, it was modified more and more by new elements which the teaching of Father Clavius had not foreseen. In fact, until the end of the 16th century, the scientific evolution of Europe had progressed usefully, but without any great brilliancy ; on the other hand, at the beginning of the 17th century, a decisive thrust forward took place : humanity was then at one of the most important turning-points of its history and was fully aware of the new way, on which it did not tarry. Germany, France and England seem to have vied in ardour to progress : if this advance is especially marked in the domain of mathematics, astronomy and physics, the triumph of the experimental method helped also to constitute mineralogy, botany took a brilliant flight, human anatomy was renewed. In China, whilst the great Christian scholars—the Doctors Paul and Leo especially—only thought of modernizing their country by translating in their totality the most famous scientific books of Europe, certain influential missionaries did not share their enthusiasm and a violent persecution cut short this exchange of ideas. At one time, one might have thought that the tradition of Father Ricci had been eclipsed. However, after some ten years, when tolerance gained the upper hand over the spiteful zenophobia of some mandarins, a small group of Jesuits, friends of Galileo,[1] of Kepler and the founders of the observational sciences,

[1] H. BERNARD, *Galilée et les jésuites des Missions d'Orient* (will appear in the *Revue des Questions scientifiques*, probably in November 1935).

resumed the work formerly outlined, with the co-operation of Paul and Leo. On the foundations laid down by Father Ricci, an encyclopædia of the notions necessary in astronomy helped to solve definitively the problem of the calendar and the ephemerides. The spirit of Father Ricci spread to Father Schall, and, through Father Verbiest, was transmitted to the members of the French Mission : by popularizing the works of Father Clavius, Ricci well deserves to be considered as the scientific initiator of modern China.

BIBLIOGRAPHY

OF WORKS MENTIONED BY THE AUTHOR

* Titles already listed in the Abridged Bibliography on page 5 ;

† Books in the Library of the old Chinese Jesuits, Peit'ang, Peiping.

†*Almagestum Claudii Ptolomæi* per Lucam Cauricum recognitum, 1528 ; 27.

†*Almagestum Ptolomei*, Venice 1496 ; 27.

†APIANI, Petri, *Cosmographicus liber*, Antwerp 1529 ; 30.

BALL, W. W. Rouse, in *The Cambridge Modern History*, Vol. V, Cambridge 1926 ; 23.

BARTOLI, *Il Giappone*, Turin 1825 ; 70.

BARTOLI, *La Cina*, Turin 1825 ; 72, 73, 76.

BERNARD, Henri, *l'Art chrétien en Chine du temps du P. Matthieu Ricci* (*Revue d'histoire des missions*, Paris 1935) ; ii, 70, 71.

BERNARD, Henri, *Aux origines du cimetière de Chala*, Tientsin 1934 ; ii, 72.

*BERNARD, Henri, *Le Frère Bento de Goes*, Tientsin 1935 ; ii, 38, 39.

BERNARD, Henri, *Galilée et les jésuites des Missions d'Orient* (*Revue des Questions scientifiques*) ; ii, 92.

BERNARD, Henri, *Matthieu Ricci et la société chinoise de son temps* ; ii, 44, 61.

*BERNARD, Henri, *Missionnaires* ; 5, 38, 39, 40.

BERNARD, Henri, *Un précurseur de la Chine moderne, Paul Siu Koang-k'i* (*Dossiers de la Commission synodale*), Peking, 1933 ; 76.

BIGOURDAN, G, *L'Astronomie*, Paris 1920 ; 45.

BIOT, *Essai sur l'histoire de l'instruction publique en Chine, et de la Corporation des Lettrés depuis les anciens temps jusqu'à nos jours*, Paris 1847 ; 19.

BOSMANS, H, S.J., *Ferdinand Verbiest, Directeur de l'observatoire de Pekin*, (1623-1688), Louvain 1912 ; 13.

BOSMANS, H., S.J., *Revue des questions scientifiques*, Louvain 1921; 13, 36.

CARRA DE VAUX, *The Legacy of Islam*, Oxford 1931 ; 10.

†CHEYNERI, Jacobi, *De Geographia*, Douai 1576 ; 30.

CHU TSAI-YU 朱載堉, *Sheng-shou-wan-nien li* 聖壽萬年曆, 1595 ; 76.

†CLAVIUS, *Astrolabium*, Rome 1593 ; 46, 52, 57.

†CLAVIUS, *Epitome arithmetica praticæ*, 1583, 1585 ; 44.

†CLAVIUS, *Euclidis elementorum libri XV*, 1574, 1589, 1591 ; 44.

CLAVIUS, *Fabrica et usus instrumenti ad horologiorum descriptionem peropportumni*, 1586 ; 47.

CLAVIUS, *Geometrica practica*, 1604 ; 69, 70.

†CLAVIUS, *Gnomonices libri VIII*, 1581 ; 44, 67,—69.

CLAVIUS, *In Sphæram Joannis de Sacrobosco Commentarius*, 1585 ; 41, 63, 69.

CLAVIUS, *Sphæra*, 1585 ; 41, 43, 49, 57, 62-3, 64, 68.

COMBARIEU, *Histoire de la musique*, Paris 1924 ; 27.

COMMANDINUS, Frederic, Edition of the work on Conic Sections of Apollonius of Perga, Bologna 1566 ; 35.

Constitutiones Societatis Jesu ; 28.

DE AVELAR, André, *Repertorio dos tempos conforme a nova reformaçao do Santo Papa Gregorio XIII*, 1585, 1590, 1594, 1602 ; 64.

DOUBLET, M., *Histoire de l'astronomie*, Paris 1922 ; 8.

DU HALDE, *A Description of the Empire of China*, London 1741, 2 vols. ; 20.

DUHEM, P., *Histoire des doctrines cosmologiques de Platon à Copernic* ; 48.

EUCLID, *see* CLAVIUS, VOEGELIN.

FINE, Oronce, *Pratica Arithmetica* ; 26.

GAILLARD, Louis, S.J., *Nankin d'alors et d'aujourd'hui, Aperçu historique et géographique*, Shanghai 1903 ; 60.

†GEMMA, Frisius, *De principiis Astronomiæ et Cosmographiæ*, Antwerp 1529 ; 30.

†GEMMA, Frisius, "Practical Arithmetic," Frankfort 1580 ; 28.

*GILES, Lionel, *Translations from the Chinese World-map of Father Ricci*, (*Geographical Journal*), 1918-19 ; 5, 38, 39, 40, 60, 62, 79.

†GLAREANUS, Henricus, *De Geographia liber*, Cologne 1581, 1591 ; 30.

GUIMARAES, Rodolph, *Les mathématiques en Portugal*, 2nd ed., Coimbra 1909 ; 38.

HAKLUYT'S *Voyages*, *Everyman's Library* 1926 ; 44.

†HARTMANN, Georges, *Perspectiva communis*, Nürnberg 1542 ; 27.

*HAVRET, Henri, S.J., *Mélanges sur la chronologie chinoise*, Coll. des *Variétés sinologiques*, No. 52, Shanghai 1920; 5, 80, 81, 82, 86.

HAY, *De rebus japonicis indicis et peruanis epistolæ recentiores*, Antwerp 1605; 50.

HEDIN, Sven, *Southern Thibet*, Stockholm 1922; 58.

HERMANN, *Buddhistische Karten*, in SVEN HEDIN'S *Southern Thibet*, VIII, 1922.

HOANG, P., *Mélanges sur l'administration*, Shanghai 1902; 73.

KAEMPFER, *The History of Japan*, Glasgow 1906, 3 vols.; 71.

LAURAND, L., *Manuel des études grecques et latines*, Paris 1930; 14.

LAVISSE ET RAMBAUD, *Histoire générale, du XVe siècle à nos jours*, Paris, 12 vols.; 23.

MASPERO, Henri, *La Chine antique*, Paris 1927; 78.

†MAUROLICUS, *Computus ecclesiasticus*; 28, 35.

†MAUROLICUS, *Opuscula mathematica*, Venice 1575; 35.

MERKEL, F. R., *G. W. von Leibniz und die China-Mission*, Leipzig 1520; 1.

MISSON, *Les idées pédagogiques de St. Ignace de Loyola*, Paris 1932; 32.

Monumenta Historica Societatis Jesu, (Work in course of publication during the last thirty years, Madrid-Rome); 26, 32.

Monumenta Ignatiana, part of the Coll. *Monumenta Historica Societatis Jesu*; 33.

Monumenta Pædagogica, Madrid 1901; 26, 29, 30, 33, 35.

Monumenta Xaveriana, special section of the Coll. *Monumenta Historica Societatis Jesu*, Vol. 1, Madrid 1899-1900; 39.

MURDOCK AND YAMAGATA, *A History of Japan*, London 1925-6, 3 vols.; 70.

NADAL, *Epistolæ P. Nadal* in the Coll. *Monumenta Historica Societatis Jesu*; 28.

PACHTLER, G. M., S.J., *Monumenta Germaniae Pædagogica*, Berlin 1887; 26, 36.

PECKHAM, "Ordinary Perspective" (*Perspectiva Communis* edited by Cardano); 28.

PELLIOT, Paul, in the *T'oung Pao*, 1934; 14.

*PFISTER, Louis, S.J., *Notices biographiques et bibliographiques de tous les membres de la Compagnie de Jésus qui ont vécu en Chine . . . depuis la mort de St. François Xavier jusqu'à la suppression de la Compagnie.* Coll. des *Variétés Sinologiques* No. 59, Shanghai 1932; 5, 43, 57, 68, 69, 70, 74, 76.

PICCOLOMINI, Alexandro, *Sfera del Mondo*; 41.

PINOT, Virgile, *La France et la formation de l'esprit philosophique en France*, Paris 1932; 77.

POLANCO, *Polanci Complementa* in the Coll. *Monumenta Historica Societatis Jesu*; 32.

†PURBACH, George, *Theoricæ novæ planetarum*, Paris 1543, Basel 1573, Cologne 1581, 1591; 15, 26, 27.

Ratio studiorum Societatis Jesu, in PACHTLER, *Monumenta . . .*, Berlin 1887; 36.

REGIOMONTANUS, Joannes, *De triangulis omnimodis* (*c.* 1464); 26.

†REGIOMONTANUS, Joannes, *Epitoma Joannes de Monte Regio in Almagestum Ptolomei*, Venice 1496; 26, 27.

†REINHOLD, Erasmus, *Prutenicæ tabulae cœlestium motuum*, Tübingen 1551, 1585; 27, 90.

REY, A., *La science orientale avant les Grecs*, Paris 1930; 82.

*RICCI, Matteo, S.J., *Opere storiche*, edited by Tacchi-Venturi, Macerata, 1910-13, 2 vols., referred to in the Notes as R.; 5, 69.

*RIQUEBOURG, *see* page 5; 40, 42, 43, 50.

RUSCELLI's Edition of Ptolemy, Venice 1561; 30.

*SAUSSURE, *Archives*; 5, 83, 89,

*SAUSSURE, *Origines*; 5, 53, 76, 78, 80, 81, 84, 85, 90.

†SCHEUTELIUS, Joannes, *Brevis regularum Algebræ descriptio, una cum demonstrationibus geometricis*, Basel 1550; 35.

†SCHONERI, Andrae, *Gnomonice*, Nürnberg 1551; 28.

†SCHONERI, Joannis, *De usu globi terrestris*, 1551; 28.

SCHURHAMMER in *Artibus Asiæ*, 1930-32; 70.

*SMITH, D. E., *History of Mathematics*, 2 vols., Boston 1923; 5.

SMITH, D. E., *Rara Arithmetica*, New York 1908; 26, 28, 44.

SMITH AND MIKAMI, *A History of Japanese Mathematics*, Chicago 1914; 14, 71.

SOMMERVOGEL, Carlos, S.J., *Bibliothèque des écrivains de la Compagnie de Jésus*, 10 vols., Bruxelles-Paris 1900 foll; 33, 41, 44, 47.

*SOUCIET; 5, 14, 51, 64, 68, 73, 76, 89.

STEVENSON, Edward Luther, *Terrestrial and Celestial Globes*, New Haven 1921, 2 vols.; 42.

†STOEFFLERINUS, Joannes, *Elucidatio fabricæ ususque Astrolabii*, Paris 1553; 26, 28.

Ta Tung Li 大統歷; 14.

†TARTAGLIA's "Algebra," Venice 1543; 24, 35.

TAYLOR, E. G. R., *A 16th century Manuscript Navigating Manual in the Society's Library*, (*Geographical Journal*), 1931; 38.

TAYLOR, E. G. R., *Tudor Geography* 1485-1583, London 1930; 38.

Transactions of the Royal Asiatic Society of Japan, Vol. 41, 1913; 71.

TRIGAULT, *De Christiana Expeditione apud Sinas*, Lyons 1616; 59.

*TRIGAULT, *Litteræ Societatis Jesu e regno Sinarum* 1610-1611, Augsbourg 1615; 5, 14, 73, 75, 76, 77, 90.

VATH, Alfons, *Johann Adam Schall von Bell*, Cologne 1933; 79.

VIVIEN DE ST. MARTIN, *Histoire de la géographie et des découvertes géographiques depuis les temps les plus reculés jusqu'à nos jours*. Avec atlas, Paris 1873; 30, 37.

†VOEGELIN, Jean, *Elementum Geometricum ex Euclidis Geometriae*, Paris 1543; 26.

*WIEGER, L., *China throughout the Ages*, Hsienhsien 1928; 5, 7, 9, 14, 76, 77, 78, 79.

WIEGER, L., *Textes historiques*, Hsienhsien; 78.

WYLIE, A., *Chinese Researches*, Shanghai 1897; 7, 11, 45.

*WYLIE, A., *Notes on Chinese Literature*, Shanghai 1902 (Publication taken over by the French Bookstore, Peiping).

Yüan Shih (History of the *Yuan*); 11, 79.

YULE, Sir H., *Travels of Marco Polo*, London 1929, 2 vols.; 59.

Yung Lo Ta-tien 永樂大典; 14.

INDEX

Page

Alfonsine Tables 15, 27, 29, 90
ALFONSO X, the Wise 15
ALGEBRA 24, 35
ALMAGEST of Ptolemy 8, 9, 15, 27
APOLLONIUS OF PERGA 19, 35
ARAB contribution to civilization 8-10
— savants, experts in method 10, 16, 17
ARABS, arithmetic of 10, 17, 18, 21
ARCHIMEDES 9, 24, 70
ARCHITECTURE, Ricci and 31, 71
—, science of European, brought to China by Ricci .. 71
ARISTOTLE 9, 16, 17, 24, 25, 34
ARITHMETIC in China 7, 18
ASTROLABE 46
ASTROLOGY 54, 65, 84
ASTRONOMER, Ricci's request for European .. 55-56
ASTRONOMICAL instruments at Nanking and Peking 11-13, 59-60
—, —, error of 13, 16
— science, Ricci and 65
ASTRONOMY, Chinese, equatorial character of .. 85, 90
—, —, intimately allied to sovereign pontificate .. 55
—, —, notions on 54, 65, 66
—, —, the problem of 53-72
—, —, reform of and Jesuit missionaries .. 74-75
—, —, reform of, obstruction by Chinese officials 75-76
—, —, weakness of 85, 86-87, 87-88
—, history of 53-54
BAROQUE 31
BEDA, Noël 25
BOARD OF MATHEMATICS 51, 61, 66, 75, 76, 79, 86
BOARD OF RITES 74, 75
BOETHIUS 27
BONCOMPAGNI, Hieronymo de 42

Page

BURNING OF THE BOOKS 78
CALENDAR, Chinese, correction of 14, 79, 80, 88
—, —, de Ursis analyses 79
—, —, logical 83
—, —, prohibition to publish a 78
—, —, reform of arrested 72, 90-91
—, —, religious importance of 54
—, Gregorian, and Ricci 55, 58, 64-65
—, Julian, reform of 55
CALVIN 25
CARDAN 24
CARTARO 42
CELESTIAL SPHERE, *hun hsiang* 渾象 11
CENTRAL OBSERVATORY OF PEKING 12, 13
CHAO-CH'ING FU 肇慶府 39, 43
CHAO CHOU 詔州 55, 66
Chien-I 簡儀, compendium instrument 12
CHINA, arithmetic in 7, 18
— becomes acquainted with Greek science 8
—, European science of architecture brought to .. 71
—, geometry intuitive in 18-19
—, Islam in 7, 10
—, Islamic bequest to 8, 10, 14
—, modern, Ricci scientific initiator of 93
—, pure mathematics in 18
—, scientific thought in 7
CHINESE astronomy, equatorial character of .. 49, 85, 90
— books, Ricci deciphers 38-39
— calendar, correction of 64, 79, 80, 88
—, —, de Ursis analyses 79-80
—, —, logical 83
—, —, reform of arrested 72, 90-91
— division of degrees and minutes 87
— ephemerides, de Ursis' conclusions *re* 88
— hour 80-81
— knowledge of equations and circle 19
— — of movements of planets 42, 89
— mathematical sciences, two classes of 78
— method of evaluating progress of sun in space .. 87
— — of fixing beginning of year 81, 82
— — of measuring time 65, 80, 82-83, 87

Page

CHINESE officials obstruct reform of calendar 75
— pupils, Ricci and 49, 57-58
— quadrant 81
—, Ricci explains Universe to 41
—, Ricci's knowledge admired by 39, 43
— science, Islamic influence on 8, 21
— —, Ricci and 39-52
— —, Ricci's high estimation of 39
—, scientific books translated into .. 45, 47, 67, 68, 69
— scientific knowledge 48
— time data and those of Europe 81
Chiu-Chang Suan-Shu 九章算術 7, 20
CHRONOLOGY, Biblical and Chinese 77-78
CH'Ü T'AI-SU 瞿太素, exaggerated estimation of Ricci 50-51
—, Ricci's disciple 44, 45, 46, 47, 50, 57, 58, 66, 69
CHUNG CHI 中氣 (Chinese Equatorial Divisions) 85, 86
CIRCLE, Chinese knowledge of the 19
CIVILIZATION, Arab contribution to 10
CLAVIUS, FATHER, mathematical knowledge spread by .. 33
—, and reform of Julian calendar 33, 55
—, Ricci's master 33, 35, 36
—, Ricci's relations with 35
—, Ricci popularizes works of 93
—, his works translated 69
CLOCK-MAKER, Ricci a 40
COCHIN 38
COLLEGE OF CHINESE MATHEMATICIANS 59
COLLEGIO ROMANO 23, 33, 35, 42, 70, 74
—, curriculum of 26-30
COLUMBUS, Christopher 27
CONGREGATIONS of Holy Virgin and Academies 32
CONSTANTINOPLE 15
COSIMO DE' MEDICI 25
COSMOGRAPHY 41
CYCLE OF METON 79
CYCLE, Sexagesimal, The 82
DE URSIS analyses Chinese calendar 79
—, Conclusions *re* Chinese ephemerides 88
—, energetic work of 76
—, mistake on ecliptic 86
—, his programme of reform 75, 76

Page

DE URSIS, successor to Ricci 74
DIOSCORIDES (1st Century A.D.) 17
DRAUGHTSMAN, Ricci a 40
ECLIPSE of sun 51, 64, 73-91
— — enhances Ricci's reputation 51
ECLIPTIC, de Ursis' mistake *re* 86
—, Ricci's mistake *re* 86
EFFECT of Ricci's teaching 63-64
ELEMENTS of Euclid, *See* Euclid
ELEMENTS, THE FIVE 17, 48, 53, 57, 84
EPHEMERIDES, Chinese, de Ursis' conclusion on .. 77, 88
—, scope of 85
EQUATIONS, Chinese knowledge of 18
EQUATORIAL character of Chinese astronomy 49, 54, 85, 90
EQUINOXES, precession of 88-89
EUCLID, Elements of 9, 18, 23, 26, 28, 34, 35, 44, 45, 67, 68
—, translation of into Chinese • 67-68
EUROPEAN architecture, science of brought to China by Ricci 71
— astronomer, Ricci's request for a 56, 66, 71
EXACT sciences neglected by the Ming 19-20
FENG YING-CHING 馮應景 61
FILIBERTO, Emmanuele 42
FIRST duty of sovereign 54, 55
FOUR ELEMENTS, Ricci's treatise on the 17, 48, 53, 57, 63, 84
GAMA, Vasco da 37
GAUBIL, Father 82, 86
GENGHIS KHAN 9
GEOMETRY in China, intuitive 18-19
GHERARDO OF CREMONA 8-9
GOA 38, 55, 66, 74
GREEK science, Chinese become acquainted with 8
GREGORIAN calendar, Ricci and the .. 55, 58, 64-65
GREGORY XIII 33
HANGCHOW 69
HEAVEN, the Vicar of 54
HIPPARCHUS 55, 89
HIPPOCRATES 22
HISTORY of Chinese astronomy 53, 54, 65
HOLY VIRGIN, Congregation of the, and Ricci 32
HOUR, the Chinese 80-81
HSING YÜN-LU 邢雲路 51

Page

HSÜ KUANG-CH'I 徐光啓, (Paul) .. 67, 69, 74, 75, 78, 90, 92
Hui-Hui Li 回回曆 14, 79
HULAGU, Grandson of Genghis Khan 9
Hun Hsiang 渾象, celestial sphere 11
IGNATIUS, Saint 28, 32
Ilkhanian Tables 9
INFLUENCE of Islam on Chinese science 8, 21
INITIATOR of modern China, Ricci was scientific 93
INSTRUMENTS, scientific 11-13, 42, 45
ISLAM in China 7
— — Europe 7, 8, 21
—, bequest to China 8, 10, 14
—, influence of on Chinese science 8, 21
JAPAN, relations with 70, 71
JESUIT group resumes Ricci's work 93
— missionaries and reform of Chinese calendar 74-75, 79
JESUS, Society of 25
JORDANUS NEMORARIUS (Le Forestier) 26
JULIAN calendar, Clavius' reform of the 55
K'AI FENG FU 60
Kao Piao 高表, gnomon 12
KEPLER 15, 49
KNOWLEDGE, Chinese, of equations and circle 19
—, mathematical, spread by Clavius 33
— of movements of planets, Chinese .. 16, 41, 42, 89
—, Ricci's admired by Chinese 43
KUBLAI 60
KUO SHOU-CHING (1231-1316) 郭守敬(若思) 9, 11, 51, 60, 79, 88, 89
LE FÈVRE D'ETAPLES (Jacobus Faber Stapulensis) .. 27
LEGACY, Islamic to China 7, 8, 10, 14
LEIBNIZ 1, 18
Li, Reduction into Chinese measures of distances .. 64
LI CH'UN 立春 82
Li Fa 曆法, Rule of Order 78, 79
LI HSIN-CHAI 李心齋 57
LI WO-TS'UN (Dr. Leo) 李我存, (Li Chih-tsao 李之藻, 振之, 涼菴) 66, 69, 75, 78, 90, 92
Ling-lung I 玲瓏儀, armillary sphere 12
LISBON 33
LOGISTICS 18

Page

LONGOBARDI 50
LUNAR year 80, 81
MA T'ANG 馬堂, the eunuch 61
MACAO 38, 39, 70
MACERATA 23
MAFFEI, Father 38
MANASE DOSAN 70
MATHEMATICAL knowledge spread by Clavius 33
— study 19, 20, 28-29, 34, 36
MATHEMATICIAN, Kuo Shu-ching 郭守敬, a .. 9, 11
MATHEMATICS 18, 34, 36, 39
MAP of World 40, 60-1, 70
MEACO (Japan) 70
MEASURING time, Chinese method of 65, 80, 82-83, 85, 86
MECHANIC, Ricci a 40
MENELAUS 35
MERAGHA (near Tabriz) 9
MERCATOR 30
MESSINA 25
METHOD, Arab savants experts in 10, 16, 17
—, Chinese, of evaluating progress of sun in space .. 87
—, —, of measuring time 82-83
MING, exact science neglected by the .. 14, 19-20, 89
MONGOL dynasty, scientific instruments constructed in .. 11
— — and Western science 8, 9 11-13
MORALES, Gabriel 28
MOVEMENTS of planets, Chinese knowledge of .. 42, 89
MOUNT SUMERU 58
MUSIC 27
MUSSULMAN BOARD OF MATHEMATICS 76, 86
MUSSULMANS, astronomy, relative perfection of their 10, 13
MUTATIONS 18
NADAL, Father 25-27
NAN CH'ANG 南昌 47, 49, 50, 51, 57, 60
NANKING 11, 13, 55, 57, 58
NASIR AL-DIN (1201-1274) 9
NATURAL HISTORY 16, 23
NICOLAS, Brother 70
NUNEZ, Pedro 38
OBSTRUCTION of Chinese officials to calendar reform 75-6
OPTICS 17, 24, 27, 40

Page

ORTELIUS 30
PAINTING, Ricci and 31
PALESTRINA 33
PANTOJA, de (Brother) 74
PASIO, Father 76
PEKING 9, 11, 12, 13, 51, 56, 59, 60, 73
PERSIAN MANUSCRIPTS 14
PERSPECTIVE, Ricci and 31
PHYSICO-MORAL DETERMINISM of Chinese Philosophy .. 89
PHYSICS 17, 25, 28, 31, 34
P'ING YANG (Shansi) 平陽 14
PLANETS, Chinese knowledge of movement of 16, 41, 42, 89
PLATO 25, 34, 48, 49
PLINY 16, 17
POLANCO 29
POSTEL, Guillaume 26
PRECESSION of equinoxes 88-89
PRISM 40
PROBLEM OF THE WEEK 82, 83
PROGRESS of sun in space, Chinese method of evaluating .. 87
PTOLEMY, his *Almagest* .. 8, 9, 15, 16, 27, 52, 55, 86
PURE mathematics in China 18
PYTHAGORAS 27
QUADRANT of circle 80-81
REFORM of Chinese astronomy 61, 74-75
— — calendar arrested 72, 90-91
REFORM, de Ursis' programme of 76
— of Julian calendar, Clavius and 33, 55
RELATIONS WITH JAPAN 70, 71
RHO, Father, and measurement of time 83, 91
RICCI, architect 31
—, arrival in Far East 7, 38
—, and astronomical science, ... 36, 65
—, builder, ... 40
—, calculates the eclipse of the sun of May 11, 1603 .. 64
—, goes to Chao-ch'ing Fu 39
—, and Chinese pupils 49, 57-58
—, and Chinese science 37-52
—, Ch'ü T'ai-su, his disciple 44, 45, 46, 47
—, Clavius and 33, 35, 41, 47
—, clockmaker 40

Page

RICCI, FATHER, congregation of Holy Virgin and Academies 32
—, constructs scientific instruments 42
—, corrects his estimate of Chinese science .. 39, 40, 52
—, death of 72
—, deciphers Chinese books 38-39
—, draughtsman 40
—, effect of his teachings 42, 63-64
—, explains Universe to the Chinese 40, 41
—, facile intellect of 35
—, and Gregorian calendar 55, 58, 64-65
—, his high estimation of Chinese science 39
—, his knowledge admired by Chinese .. 39, 43
—, his Map of the World 40, 50, 60, 61, 62
—, his memory of locality 35
—, measures the longitude of Peking 64
—, mechanic 40
—, mistake on ecliptic 86
—, modesty 36
—, at Nanking 11, 58-59
—, and painting 31
—, and perspective 31
—, popularizer of Clavius' works 67, 93
—, qualifications of 36, 39-41
—, requests European astronomer be sent to China 55-56, 71
—, results of his teaching 45, 71
—, at Rome 23
—, scientific books seized 60-61
—, scientific initiator of modern China 93
—, scientific training of 22-36
—, second visit to Peking 56, 60
—, Small Treatise on the Error of the Five Elements, 57; Reprinted 61
—, his work resumed by Jesuit group 93
RUGGIERI, Father 38, 39
SACROBOSCO, Johannes de 15, 63, 69
SAINT IGNATIUS OF LOYOLA 25, 27, 32
SAVANTS, Arab, experts in method 10, 16, 17
SCHALL and measurement of time 83, 91
— and Chinese Board of Mathematics 79
SCIENCE, astronomical, Ricci and 36, 65
—, Chinese, Islamic influence on 8, 21

Page

SCIENCE, Chinese, Ricci and 36, 37-52
—, —, Ricci's estimate of 39, 40
—, Greek, Chinese become acquainted with 8
—, Western, Mongol dynasty and 9-10
SCIENCES, exact, neglected by Ming 14, 19-20
—, mathematical, two classes of Chinese 78
SCIENTIFIC acquirements, Ricci's 38, 48
— books seized by Chinese 60-61
— — translated into Chinese 67, 68, 69
— initiator of modern China, Ricci was 93
— instruments 11-13, 42, 45
— — constructed in Mongol dynasty 11
— —, Ricci constructs 42
— knowledge, Chinese 48
— method and tradition 23-24
— reputation, Ricci's 48
— thought in China 7
— training, Ricci's 23
SILK 78
SIXTUS V 33
SEXAGESIMAL Cycle 82
Shou-Shih-li 授時歷 79
SOCIETY OF JESUS 25, 26, 28, 29, 36
SOLAR eclipse of December 15, 1610 73-91
SOOCHOW 51, 60, 69
SOVEREIGN, first duty of 54, 55
SPACE, Sun in, Chinese method of evaluating progress of .. 87
SPHERE, celestial 11
Su 宿 82, 83, 84, 85
SUCCESSOR, de Ursis is Ricci's 74
SUN, Eclipse of enhances Ricci's reputation 51
— in Space, Chinese method of evaluating progress of 87
SUPERSTITIOUS ACT of calculating the calendar 79
Ta T'ung-li 大統歷 51
TEACHING, effect of Ricci's 42, 63-64
TERRENZ, Father, and measurement of time .. 83, 91
THEODOSIUS 35, 69
THOUGHT, scientific, in China 7
TIENTSIN 61
T'ien Wen 天文, "The Ornament of Heaven" .. 78, 79
TIME, Chinese method of measuring 65, 82-83

Page

TIME data, Chinese and European 81-82
TIMOCHARIS 89
TOLEDO 8
TORRES, D. 28, 29
TRANSLATION of scientific books into Chinese .. 67, 68, 69
TRIGONOMETRY 19, 24
TYCHO BRAHE (1546-1601) 11, 13
ULUGH BEG (1393-1449) 14
URANIBORG (Observatory) 11
UNIVERSE, Ricci explains to the Chinese 40, 41
URSIS, *see also* de Ursis 74-80, 81-92
VALIGNANO, Father 39, 52
VERBIEST, Father, and Chinese Board of Mathematics 13, 79
—, Ricci's spirit spread by 93
VICAR of Heaven 54
VIGNOLA, Architect of the Gesù (1568) 31-32
VISIT to Peking, Ricci's second 60
VITELO 27
WEAKNESS of Chinese astronomy 85-88
WESTERN science, Mongol dynasty and 9-10
WORK, Ricci's resumed by Jesuit group 93
WORKS of Clavius translated into Chinese 69
— —, Ricci popularizer of 93
WORLD, map of 40, 50, 60, 61
WYLIE, A. 5, 13
WU, the Emperor 78
WU TI, the Emperor 88
WU TSO-HAI 吳左海 60
Yang 陽 53, 84
YAO, the Emperor 78
YEAR, Chinese method of fixing beginning of .. 81, 82
—, lunar 80, 81
YEH-LÜ CH'U-T'SAI 耶律楚才 9, 13
Yin 陰 53, 84
Yin 寅 82
YULE, Col. H. 13